SpringerBriefs in Architectural Design and Technology

Series Editor

Thomas Schröpfer, Architecture and Sustainable Design, Singapore University of Technology and Design, Singapore, Singapore

W0037687

Indexed by SCOPUS Understanding the complex relationship between design and technology is increasingly critical to the field of Architecture. The *Springer Briefs in Architectural Design and Technology* series provides accessible and comprehensive guides for all aspects of current architectural design relating to advances in technology including material science, material technology, structure and form, environmental strategies, building performance and energy, computer simulation and modeling, digital fabrication, and advanced building processes. The series features leading international experts from academia and practice who provide in-depth knowledge on all aspects of integrating architectural design with technical and environmental building solutions towards the challenges of a better world. Provocative and inspirational, each volume in the Series aims to stimulate theoretical and creative advances and question the outcome of technical innovations as well as the far-reaching social, cultural, and environmental challenges that present themselves to architectural design today. Each brief asks why things are as they are, traces the latest trends and provides penetrating, insightful and in-depth views of current topics of architectural design. *Springer Briefs in Architectural Design and Technology* provides must-have, cutting-edge content that becomes an essential reference for academics, practitioners, and students of Architecture worldwide.

Sanja Stevanović

Overhang Design Methods

Optimal Thermal and Daylighting Performance

 Springer

Sanja Stevanović
Mathematical Institute of the Serbian
Academy of Sciences and Arts
Belgrade, Serbia

ISSN 2199-580X   ISSN 2199-5818 (electronic)
SpringerBriefs in Architectural Design and Technology
ISBN 978-981-19-3011-9   ISBN 978-981-19-3012-6 (eBook)
https://doi.org/10.1007/978-981-19-3012-6

This Springer imprint is published by the registered company Springer Nature Singapore Pte Ltd.
The registered company address is: 152 Beach Road, #21-01/04 Gateway East, Singapore 189721,
Singapore

To Milica, Djordje and Dragan

Contents

Chapter 1
Introduction

Abstract Modern façades, in particular those of office buildings, are highly glazed both to provide daylight and exterior views, which influence occupants' well-being, and to improve the architectural appearance of buildings. Solar radiation through such façades easily causes glare and overheating, so shading is of vital concern for them. Shading devices aim to prevent excessive heat gains from solar radiation in hot periods, without compromising beneficial heat gains in cold periods or visual comfort in adjacent indoor spaces. Balancing these often contradictory goals is an interesting multi-objective optimisation problem that represents a very active research topic in the field of building energy and daylighting. Overhangs are the simplest and most traditional shading devices that have several beneficial traits: easy installation, high cost-effectiveness, low maintenance and unobstructed view outside. Our goal here is to provide an overview of historical development of the methods for designing overhangs. These methods improved hand in hand with the development of methods for obtaining information about thermal and daylighting behaviour of building models, so this book also represents an overview of the development of building performance simulation tools as related to shading calculations.

Keywords Shading · Overhang · Windows · Buildings

Overhang is usually a fixed surface, mostly horizontal or horizontally inclined, protruding above the window. It is often accompanied by side fins, mostly vertical surfaces next to the window or vertically inclined towards the window. Overhangs are usually chosen to protect equator-facing windows, while side fins are usually chosen for windows on the remaining façades. The overhang may also be adjoined by a horizontal or horizontally inclined surface projecting below the window, to protect it from the ground reflected solar radiation as well. In such cases, and especially when they serve to shade a whole array of windows, this combination of shading surfaces is called egg-crate shading. While overhangs are usually cantilever additions to the wall nowadays, they traditionally appeared in a variety of forms such as roof or wall extensions, jettied storeys, balconies, verandas or porticoes. Such fixed shading devices are a prominent feature of vernacular architecture [7], where they are mostly used to shade external walls. However, improvements in insulation properties of walls exposed windows as critical elements of building façades. It is

S. Stevanović, *Overhang Design Methods*,
SpringerBriefs in Architectural Design and Technology,
https://doi.org/10.1007/978-981-19-3012-6_1

Fig. 1.1 Examples of overhanging shading structures. Upper left: Overhangs at a building in Hong Kong Polytechnic University campus [12]; Lower left: Egg-crate shading on a western facade of a public building in Niš, Serbia; author's photograph; Right: Shading by balconies in Casa Atlântica, Rio de Janeiro, designed by Zaha Hadid Architects, adapted from a render [31]

estimated nowadays that up to half of heating load can be attributed to heat losses through windows [3, 4]. Hence, most current overhang research is about shading windows. Some examples of overhanging shading structures are shown in Fig. 1.1.

Window shading has been a topic of quite a few surveys so far. Kirimtat et al. [14] give an overview of different types of shading devices and review the main simulation tools used for their modelling in building energy and daylighting simulations. They observe that office buildings are the most common case study buildings in shading simulations, while the most commonly simulated shading devices are Venetian blinds, external fixed louvers, roller shades and overhangs. Kuhn [17] gives a comparative evaluation of both traditional and advanced solar shading devices, together with development guidelines for new devices based on his own experiences as an inventor of several innovative shading systems. Bellia et al. [5] give a very brief overview of effects of different shading devices on lighting and thermal performance of buildings, while Sharda and Kumar [25] review experimental, analytical and theoretical studies that analyse heat transfer through glazing systems with shading devices.

It is generally agreed that passive strategies should be considered early in the design of energy-efficient buildings, prior to other strategies that might unnecessarily consume energy [18, 21]. Sections on fixed solar shading, as one of the passive strategies, are thus contained in recent reviews of passive measures for energy conservations in buildings [2, 10] and reviews of passive cooling strategies [22, 27]. Prieto et al. [22] provide an estimate that shading may save on average around 25% of cooling demand in office buildings in warm–dry and warm–humid climates. Wong [29] provides an overview of research on modern light guiding systems that also serve as shading devices by protecting from excessive solar heat gain without reducing transmission of diffuse radiation within his review of daylighting design for buildings. Photovoltaic cells can be integrated into shading devices as well, as they can convert unwanted solar radiation into electricity in place [28]. Zhang et al. [33] give a comprehensive review of shading devices with integrated photovoltaics, focusing on their types, optimum tilt angle and orientation, suitable photovoltaic materials and list a number of implementations of such shading devices in real buildings. Ibraheem et al. [13] give another review of shading devices with integrated photovoltaics from the systems theory standpoint.

Many modern shading devices are automated using sensors and actuators that react to changing environmental conditions and try to optimise shading performance. In a somewhat older review, Colaco et al. [6] review the research on the influence of glazing size and orientation, automated window blinds and dimming of artificial lights by daylight sensors on energy consumption in commercial buildings. They state that it can be assumed that building occupants will not use the blinds to optimise daylight quality criteria and energy consumption, and that they will alter the shading position only when they are exposed to extreme discomfort, justifying the need for automatic control of the blind position. This assumption may be explained by the natural tendency of people to focus the use of their own vital energy on the most important goals, where usually extreme discomfort is needed to give importance to shading among everyday tasks. Konstantoglou and Tsangrassoulis [16] focus on dynamic systems that aim to keep a balance between protection from direct solar radiation and glare, and provision of sufficient daylighting levels. Most of such systems are motorised blinds, and Konstantoglou and Tsangrassoulis review both their control methods and implications for energy savings. Silva et al. [26] consider 12 different blind control strategies in office buildings and observe that they have a significantly different impact on overall energy demand for heating, cooling and lighting, concluding that different choices of blind control strategy will result in different choices of best design alternatives in building performance simulations. Further in this direction, Fiorito et al. [9] collect and review examples of biomimetic shape morphing shading systems that close or open by relying on shape memory materials, capable of reversibly varying their shape under mechanical, chemical or electrical stimuli triggered by solar radiation. Dakheel and Aoul [7] give another review of biomimetic shading systems and in addition review dynamic glazing which, with its optical properties changing under external influences, is also considered to be a form of window shading. According to them, electrochromic glazing is the most applied dynamic glazing.

Human behaviour is an influential factor in the design and operation of energy-efficient buildings. D'Oca et al. [8] draw attention to the fact that the energy-related behaviour of key stakeholders (building designers and owners, occupants, operators and managers, energy providers and policy makers) over the building life cycle is as important as the influence of technological advances. The problem, however, lies in a current lack of methods that can reliably predict the influence of human behaviour, leading to discrepancies between predicted and measured building energy performance. O'Brien et al. [20] review 12 major studies of manual window shade operation patterns in office buildings and classify factors that influence the decision to operate the shade. They observe that occupants move their shades very infrequently (weekly, monthly, seasonally and even never), primarily based on long-term solar geometry and radiation intensity rather than short-term changes, generally aiming to improve visual comfort rather than thermal comfort. The infrequency of shade operation actually suggests that assuming a fixed shade position with a realistic mean shade occlusion may lead to more accurate results in building performance simulations [20]. Wymelenberg [30] reviews behaviour specifically related to blind use, indicating numerous factors affecting occupants' use of blinds which may be not only physical (daylight, glare, direct solar radiation) but also physiological (individual sensitivity to brightness), psychological (desire for privacy or for access to view) or social (sense of blind ownership). He concludes that the way people use blinds appears to be extremely personal and that an easily definable model of blind use may not exist. Moreover, Rubin et al. [24] demonstrated that people do put a significant mental effort into choosing blind positions: 80% of the tested occupants returned their blinds to the originally chosen position within a single day after they were perturbed by researchers. Hence, it is no wonder that automated shading control methods often fail to recognise when conditions become undesirable or even move shades to a less desirable position, for which occupants have low tolerance [23]. Zhang et al. [32] focus on the current understanding and quantitative modelling of occupant behaviour in building energy performance. Their literature review shows that changes in occupant behaviour have the potential to save between 10 and 25% of energy in residential buildings and between 5 and 30% of energy in commercial buildings.

Refurbishment of existing buildings is of particular importance among efforts to reduce greenhouse gas emissions, due to the low annual replacement rate of existing buildings of only 0.07% [11]. Li et al. [19] review suitable sustainable refurbishment options for high-rise residential buildings in subtropical high-density cities such as Hong Kong. In a situation quite opposite from office buildings, there is no need to cater for sufficient daylighting or automated shading when most family members leave for work or school during the daytime, so Li et al. [19] suggest only overhangs and vertical fins as appropriate refurbishment options for shading in these cases. Finally, in their review of three types of shading systems for tropical office buildings, passive (fixed and manually adjustable), active (whose actuators use grid energy) and hybrid (based on biomimetics and properties of shape memory materials), Al-Masrani et al. [1] conclude that fixed shading devices, particularly conventional models, are most studied and used due to their simple design, easy installation, high cost-effectiveness and low maintenance. They also note that most of the studies considered in their review identify egg-crate devices as the best shading type to improve daylight and thermal performance in the tropics.

From this overview of previous surveys on shading windows of buildings, we can see that we still have a long way to go before we may understand subtle nuances of interaction between excessive solar radiation, shading devices and indoor thermal and visual comfort. In order to provide a hopefully valuable resource for this understanding, we focus the present survey on the historical development of methods for designing the simplest and most traditional fixed shading device—an overhang. While there are certainly better shading alternatives to overhangs in cases when cooling energy is predominant, such as in office buildings with high windows-to-wall ratios located in hot climates [15], overhangs still represent a viable shading option in many situations.

A historical overview of design methods based on solar positions and shading masks is given in Chap. 2. Current methodology is reviewed in Chap. 3, which contains sections on building performance simulation methods, discussion of various objective functions used to judge the optimality of shading and the use of multi-objective optimisation approaches. Chapter 4 reviews design methods for particular overhang types. Besides the optimisation studies of standard overhangs, it also discusses the approaches that break away from rectangular-shaped overhangs and suggests how overhangs should be shaped for optimal performance. In addition, it discusses the studies of using overhangs in building retrofits, integrating photovoltaic cells in overhangs, and of dynamic overhangs. Finally, the appendix classifies the overhang case studies according to climate type of their locations, as the climate type mostly governs the amount of available direct and diffuse solar radiation. Studying results obtained in similar climates may therefore help to get a general impression of the design and capabilities of overhangs and side fins in such climates.

References

1. Al-Masrani SM, Al-Obaidi KM, Zalin NA, Isma MIA (2018) Design optimisation of solar shading systems for tropical office buildings: challenges and future trends. Sol Energy 170:849–872
2. Amirifard F, Sharif SA, Nasiri F (2018) Application of passive measures for energy conservation in buildings - a review. Adv Build Energ Res. https://doi.org/10.1080/17512549.2018.1488617
3. Apte J, Arasteh D, Yu JH (2003) Future advanced windows for zero-energy homes. ASHRAE Trans 109:871–882

4. Arasteh DK, Goudey H, Yu JH, Kohler C, Mitchell R (2007) Performance criteria for residential zero energy windows. ASHRAE Trans 113(1):176–185
5. Bellia L, Marino C, Minichiello F, Pedace A (2014) An overview of solar shading systems for buildings. Energ Procedia 62:309–317
6. Colaco SG, Kurian CP, George VI, Colaco AM (2008) Prospective techniques of effective daylight harvesting in commercial buildings by employing window glazing, dynamic shading devices and dimming control–A literature review. Build Simul-China 1:279–289
7. Dakheel JA, Aoul KT (2017) Building applications, opportunities and challenges of active shading systems: a state-of-the-art review. Energies 10:1672
8. D'Oca S, Hong T, Langevin J (2018) The human dimensions of energy use in buildings: a review. Renew Sust Energ Rev 81:731–742
9. Fiorito F, Sauchelli M, Arroyo D, Pesenti M, Imperadori M, Masera G, Ranzi G (2016) Shape morphing solar shadings: a review. Renew Sust Energ Rev 55:863–884
10. Gupta N, Tiwari GN (2016) Review of passive heating/cooling systems of buildings. Energy Sci Eng 4:305–333
11. Hartless R (2003) Application of energy performance regulations to existing buildings. Final Report of the Task B4, ENPER TEBUC, SAVE 4.1031/C/00-018/2000, Building Research Establishment, Watford
12. Huang Y, Niu Jl, Chung Tm (2012) Energy and carbon emission payback analysis for energy-efficient retrofitting in buildings—Overhang shading option. Energ Build 44:94–103
13. Ibraheem Y, Farr ERP, Piroozfar PAE (2017) Embedding passive intelligence into building envelopes: a review of the state-of-the-art in integrated photovoltaic shading devices. Energ Procedia 111:964–973
14. Kirimtat A, Koyunbaba BK, Chatzikonstantiou I, Sariyildiz S (2016) Review of simulation modeling for shading devices in buildings. Renew Sust Energ Rev 53:23–49
15. Koç SG, Kalfa SM (2021) The effects of shading devices on office building energy performance in Mediterranean climate regions. J Build Eng 44:102653
16. Konstantoglou M, Tsangrassoulis A (2016) Dynamic operation of daylighting and shading systems: a literature review. Renew Sust Energ Rev 60:268–283
17. Kuhn TE (2017) State of the art of advanced solar control devices for buildings. Sol Energy 154:112–133
18. Lechner NM (2015) Heating, cooling, lighting: sustainable design methods for architects. Wiley, Hoboken
19. Li J, Ng ST, Skitmore M (2017) Review of low-carbon refurbishment solutions for residential buildings with particular reference to multi-story buildings in Hong Kong. Renew Sust Energ Rev 73:393–407
20. O'Brien W, Kapsis K, Athienitis AK (2013) Manually-operated window shade patterns in office buildings: a critical review. Build Environ 60:319–338
21. Olgyay V (1963) Design with climate: bioclimatic approach to architectural regionalism. Princeton University Press, Princeton
22. Prieto A, Knaack U, Auer T, Klein T (2018) Passive cooling & climate responsive façade design: Exploring the limits of passive cooling strategies to improve the performance of commercial buildings in warm climates. Energ Build 175:30–47
23. Reinhart CF, Voss K (2003) Monitoring manual control of electric lighting and blinds. Lighting Res Technol 35:243–258
24. Rubin AI, Collins BL, Tibbott RL (1978) Window blinds as a potential energy saver—A case study. National Bureau of Standards Building science series 112. US Government printing office, Washington
25. Sharda A, Kumar S (2014) Heat transfer through glazing systems with inter-pane shading devices: a review. Energ Tech Policy 1:23–34
26. Silva PCd, Leal V, Andersen M (2012) Influence of shading control patterns on the energy assessment of office spaces. Energ Build 50:35–48
27. Valladares-Rendón LG, Schmid G, Lo SL (2017) Review on energy savings by solar control techniques and optimal building orientation for the strategic placement of façade shading systems. Energ Build 140:458–479

28. Yoo SH, Lee ET, Lee JK (1998) Building integrated photovoltaics: a Korean case study. Sol Energy 64:151–161
29. Wong IL (2017) A review of daylighting design and implementation in buildings. Renew Sust Energ Rev 74:959–968
30. Wymelenberg KVD (2012) Patterns of occupant interaction with window blinds: a literature review. Energ Build 51:165–176
31. Zaha Hadid Architects. A render of Casa Atlântica in Rio de Janeiro, Brazil. www.zaha-hadid. com/architecture/casa-atlantica/. Accessed Feb 2022
32. Zhang Y, Bai X, Mills FP, Pezzey JCV (2018) Rethinking the role of occupant behavior in building energy performance: a review. Energ Build 172:279–294
33. Zhang X, Lau SK, Lau SSY, Zhao Y (2018) Photovoltaic integrated shading devices (PVSDs): a review. Sol Energy 170:947–968

Chapter 2
Solar Path Methods

Abstract The oldest methods for overhang design were proposed in the second half of the twentieth century and based solely on solar paths. Their aim is to protect glazing from direct solar radiation during overheating periods, usually specified by cut-off dates and cut-off times during these days. In the first two sections, we review Olgyays' and Mazria's methods based on superimposing overhang and fins' shading masks upon charts of solar paths in horizontal and cylindrical projections, respectively. In the third section, we review methods based on backward tracing of solar rays until they intersect a predefined shading support surface.

Keywords Solar paths · Horizontal projection · Cylindrical projection · Backward tracing.

2.1 Olgyays' Horizontal Projection Method

The oldest known overhang design method, developed by twin brothers Victor and Aladar Olgyay, is described in Victor's classic 1957 [22] and 1963 [23, Chap. VII] books, while according to Denzer [6], the method was introduced even a bit earlier in their joint 1954 report entitled *Application of Climatic Data to House Design*.

The first step of the method is to determine the times when shading is needed based on the average hourly temperatures for each hour and each month at the building location. Note that this period of overheating is defined based on outdoor temperatures, regardless of building properties. This step is usually simplified by replacing a detailed determination of the period of overheating with a choice of cut-off dates and times during which shading is needed. As an illustrative example, for a location of Niš, Serbia (43.32 °N, 21.93 °E), we can choose June 4–August 20 from 11 am to 4 pm as cut-off dates and times, according to average daily temperatures. One should, however, immediately notice that the solar chart is symmetric with respect to solstices in the sense that for each day between a winter and a summer solstice there is a corresponding day between a summer and a winter solstice when the Sun follows the exact same path, so that any fixed shading device which shades the window on one date will necessarily provide the same shading pattern on the corresponding symmetric date as well. This represents a slight drawback of fixed shading devices as climatic seasons tend to have a time lag of about one month after their corresponding solstices and equinoxes, due to the heat-retaining capacity of the Earth. As August 20 is symmetric with April 22 with respect to solstices, an overhang will provide the same shading for both dates, regardless of the fact that air temperatures on April 22 are about $10°$ lower than on August 20. Thus, the above choice of cut-off dates of June 4–August 20 is, from the solar chart point of view, equivalent to the choice of April 22–August 20.

One should also be precautious of the predictions of future climate changes when choosing the cut-off dates. While the climate change effects will vary with region, season and time of day, rising global temperatures imply that warm days will occur earlier and persist longer during the year, thus prolonging the shading season. For example, according to [7] the shading season in southern Europe will, on average, start and end 38 days earlier and later, respectively, than nowadays, which will lower the average maximum solar altitude by about $13.5°$.

The second step is to determine the position of the Sun when shading is needed, which is done by representing this time period on a solar chart, obtained by projecting solar paths from sky dome to the horizontal plane. The resulting projection is a circular chart with altitude angles denoted by concentric circles and azimuth angles denoted by radii. Recall that solar altitude is the angle between the solar rays and horizontal plane (from $0°$ to $90°$), while solar azimuth is the angle between the direction of North and projection of the solar rays on the horizontal plane, measured clockwise (from $0°$ to $360°$). There are several ways of representing sky dome in the horizontal plane, as noted by Szokolay [30, Sect. 2.2]: *equidistant* representation, with solar altitude circles evenly spaced in the horizontal plane, *orthographic* projection, which is rarely used as low altitude circles are mapped closely together, while high altitude circles are widely spaced, and *stereographic* projection, which introduces an additional step in the projection to obtain more evenly spaced altitude circles. Older Refs. [21, 30] mention two computer programs, written specifically to illustrate and teach Olgyays' method, but these programs cannot be found online nowadays. Recent book [24] suggests several current methods for obtaining solar charts: generating them online using [32], which produces equidistant charts with North at the bottom, or using [16], whose charts cannot be saved on a local disc, downloading them from [12] in stereographic projection for whole numbered latitudes, or generating them offline using the Solarbeam application [18]. Figure 2.1 shows solar path chart in equidistant representation created for selected location with Solarbeam [18].

The next step is to create a shading mask for a shading device that will block incidence of direct solar radiation on the point of observation during the overheated period. This construction is based on the observation that (infinite) overhangs, when mapped to sky dome and then projected to the horizontal plane, produce segmental shading masks, while (infinite) fins produce radial shading masks. The shading mask of any shading device is the union of shading masks of its basic constituents, enabling Olgyays to create the shading protractor, a visual tool for constructing shading masks of arbitrary shading devices. Figure 2.2 shows an image of an interactive

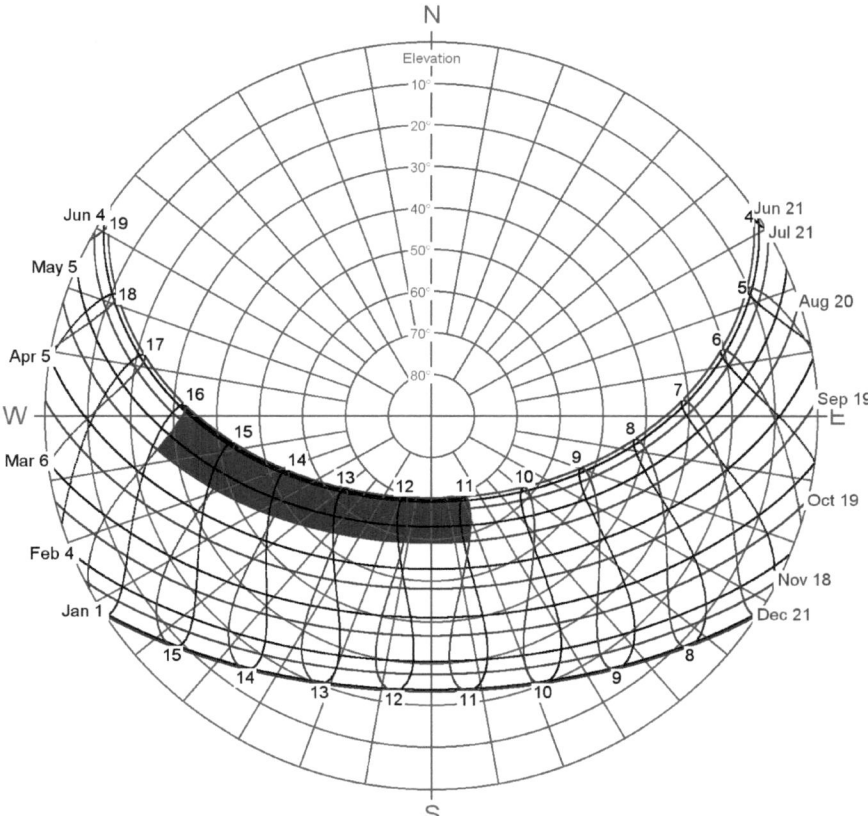

Fig. 2.1 Solar path chart in equidistant representation for a location of Niš, Serbia (43.32 °N, 21.93 °E), created with Solarbeam [18]. Shaded region represents the selected cut-off date/time choice of June 4–August 20 from 11 am to 4 pm

version of the shading protractor [27], recreated in GeoGebra following the procedure from [30]. Here, the centre of the semicircle denotes the point of observation, while the semicircle denotes the field of view outward from the point of observation, hence a half of sky dome, projected to the horizontal plane. Shading masks for many types of compound shading devices have been studied in detail by El-Refaie [9]. Note that the shading mask depends only on the angle between the normal to the window plane and the shading element, with vertex at the point of observation, i.e., on the ratios of dimensions of shading device and window elements. Hence, shading protractor enables the construction of shading masks not only for actual shading devices, but also for deep window reveals and neighbouring buildings and obstructions by measuring the angles at which they are visible from the point of observation.

In order to design an appropriate shading device, the shading protractor is then superimposed on the solar path chart and a shading mask is chosen so that it covers the

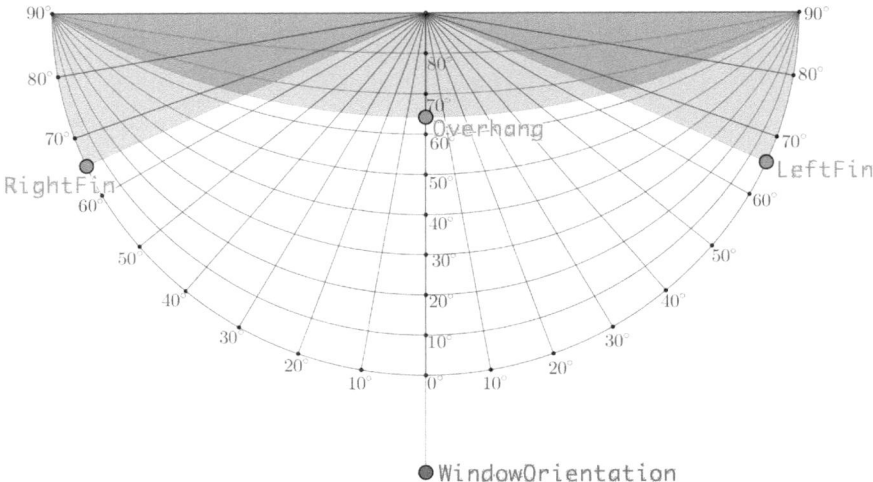

Fig. 2.2 Shading protractor recreated in GeoGebra [27]

period when shading is needed. Figure 2.3 illustrates the case of the shading protractor for a window with an azimuth of 200° (clockwise from the North) superimposed on the solar path chart from Fig. 2.1. It can be seen from this figure that the point of observation will be shaded during the requested cut-off dates and times if the outer edge of the overhang is visible at an altitude of approximately 47°, while the fins are not relevant in this case. As Olgyays [23, p. 74] noted already, overhangs efficiently protect southern orientations (apparently referring to conterminous US latitudes), while fins work well for eastern and western orientations.

A remaining choice for the use of Olgyays' method is the position of the point of observation. If the window is to be fully shaded during the cut-off dates and times, the point of observation should be placed at the bottom of the window, while if the window is to be 50% shaded the point of observation should be placed in the centre of the window. This choice has consequences on the actual dimensions of the overhang, as the overhang for 100% shading would be twice as large as the one ensuring 50% shading, which is the probable reason why Olgyays [23, p. 82/83] suggests as a rule of thumb that "the shading device will work well" if it ensures 50% shading.

Assuming that the overhang has to be designed for a window that is $w = 2$ m wide and $h = 1.4$ m high, the choice of 50% shading and the angle $\alpha = 47°$, at which the outer edge of overhang is visible from the point of observation (see Fig. 2.4), yield that the overhang should have depth

$$d = \frac{50\% \cdot h}{\tan \alpha} = 0.65 \, \text{m}.$$

Note that the shading protractor assumes the overhang to have infinite length, to avoid considering the actual shape of the shadow projected on the window. A realistic

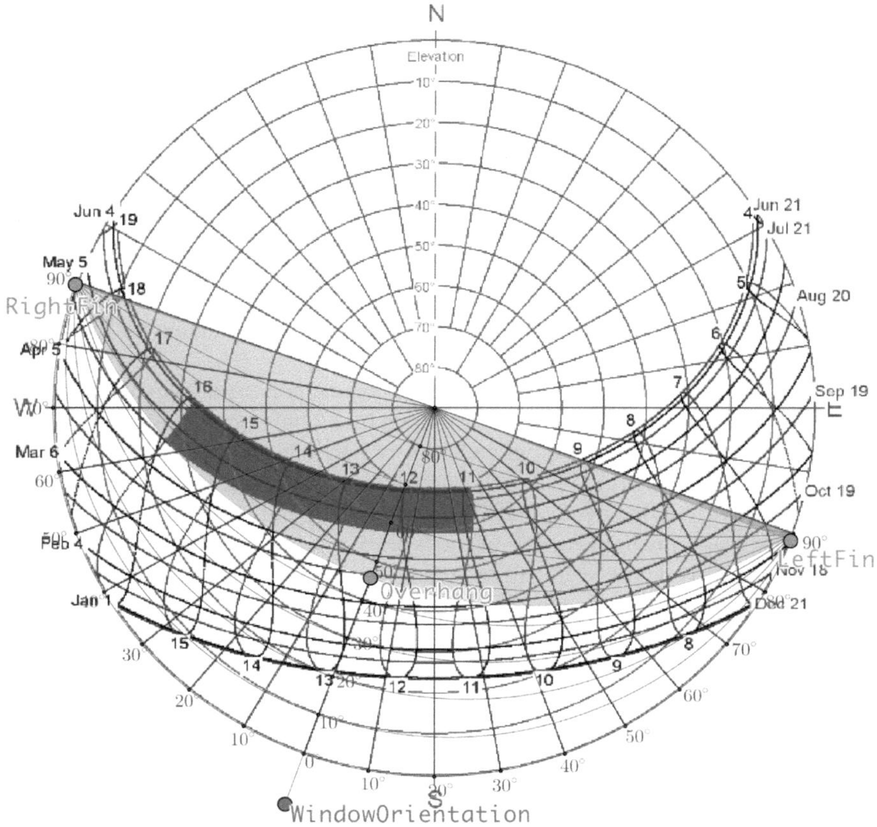

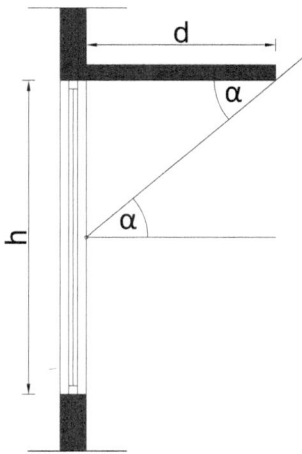

Fig. 2.3 Shading protractor superimposed on a solar path chart

Fig. 2.4 Vertical section of the overhang and the window

overhang should thus either extend sideways sufficiently long or be combined with appropriate fins to ensure shading of the point of observation during whole cut-off dates and times.

As Leatherbarrow and Wesley already observed [15], the shading protractor has been since long replaced by computer simulations. However, due to the simplicity of the method and the large influence of Olgyay's book [23], examples of its use can still be found in recent shading studies: Subhashini and Thirumaran [29] used it to suggest slanted vertical fins as appropriate shading devices for a classroom in the warm humid climate of Madurai, India; Tahbaz [31] showcased the use of Olgyays' method to design shading devices for the central Iranian city of Bam with a hot arid climate, while Arias [1] used it to discuss shading requirements for the humid subtropical climate of Guadalajara in Mexico.

Etzion [10] called attention to inherent imprecision present in Olgyays' method: it indicates whether the window is lit or shaded based only on whether the point of observation is exposed to direct solar radiation or not. When the point of observation is at the bottom of the window, it will happen in borderline cases where the point of observation is lit while overhang still shades most of the window. Even when the point of observation is in the centre of the window, it may happen that the point of observation is lit while the overhang and a fin still shade almost three-quarters of the window. Etzion [10] thus proposed an extension of Olgyays' method that serves to visually determine the proportion of the shaded part of the window at any given time. However, this extension works for a single chosen time only and cannot be used to visualise the gradual transition from a fully shaded to a fully lit window on a solar path chart.

2.2   Mazria's Cylindrical Projection Method

Edward Mazria in his 1979 book [19] proposed an overhang design method that is fully analogous to that of Olgyays, with the way of projecting solar paths being the only difference. Namely, instead of projecting solar paths to a horizontal plane, Mazria suggested projecting them to a vertical cylindrical surface surrounding the point of observation, which is then unwound into a rectangular chart. More specifically, solar paths are plotted on a rectangular chart in which horizontal coordinates represent solar azimuths, while vertical coordinates represent solar altitudes. Figure 2.5 shows the solar path chart in this projection created with the online tool [33] for the location of Niš, Serbia, with the shaded region corresponding to cut-off dates of June 4–August 20 and cut-off times from 11 am to 4 pm. One more way of imagining the process of obtaining this solar chart is to cut Olgyays' circular solar chart along the segment connecting North with the chart centre and to straighten the chart's perimeter. Since the result of this straightening should be a rectangle (and not a triangle), this inevitably leads to a large amount of distortion at higher solar altitudes—for example, a single zenith point corresponds to the whole upper side of the chart representing the solar altitude of 90°.

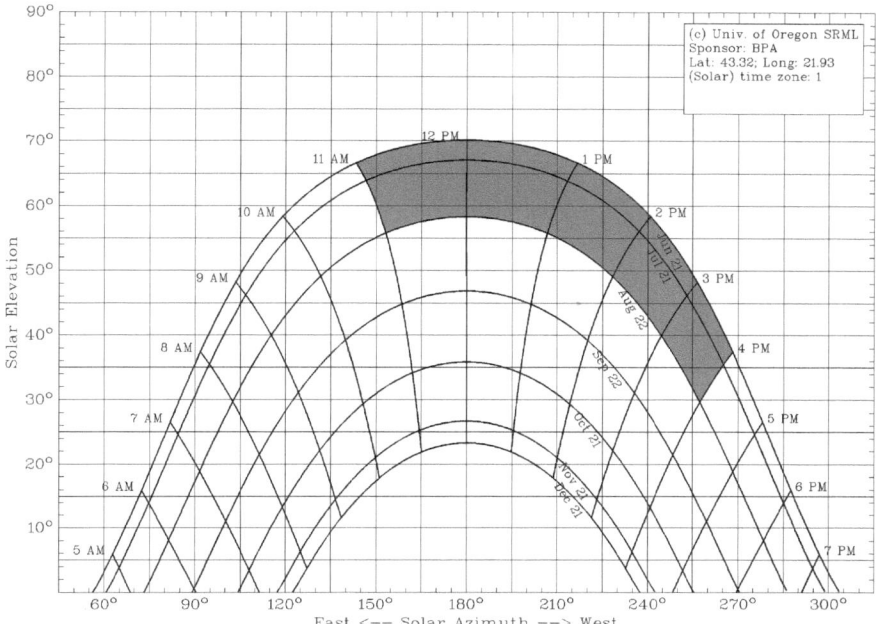

Fig. 2.5 Solar path chart in cylindrical projection for a location of Niš, Serbia (43.32 °N, 21.93 °E), created with the online tool [33]. Shaded region represents the selected cut-off date/time choice of June 4–August 20 from 11 am to 4 pm

This change in the projection of solar paths is also reflected in the different shapes of the shading protractor. In the cylindrical projection, the outer edge of the overhang is mapped to a parametric curve given by

$$\gamma = \arctan\left(\tan\alpha\cos\beta\right), \tag{2.1}$$

where α, β and γ are angles with vertices at the point of observation: α is the angle at which the outer edge is visible, $0° \le \beta \le 90°$ is the angle between the projection of a point on the outer edge to a horizontal plane and a normal to the window plane, while $0° \le \gamma \le 90°$ is the angle between the point on the outer edge and the horizontal plane (see Fig. 2.6).

The shading mask of the (infinite) overhang is the area above this curve, while shading masks of (infinite) fins are rectangular areas between the fin positions and the ends of the protractor. Figure 2.7 shows an image of an interactive version of the shading protractor for Mazria's projection [28], recreated in GeoGebra.

As in Olgyays' method, the shading protractor is superimposed on the solar path chart, aligning the protractor's pointer (labelled with 0°) with window orientation and aiming for a shading mask to cover the overheating period. As in the previous section, Fig. 2.8 illustrates the case of the shading protractor for a window with

Fig. 2.6 Visual
representation of the angles
α, β and γ from Eq. (2.1)

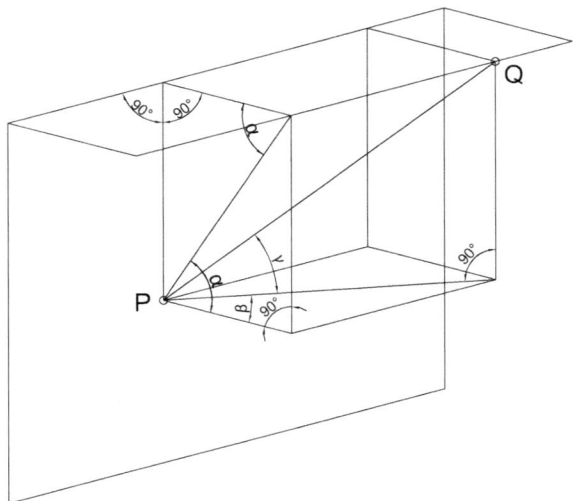

Fig. 2.7 Shading protractor
for cylindrical projection,
recreated in GeoGebra [28]

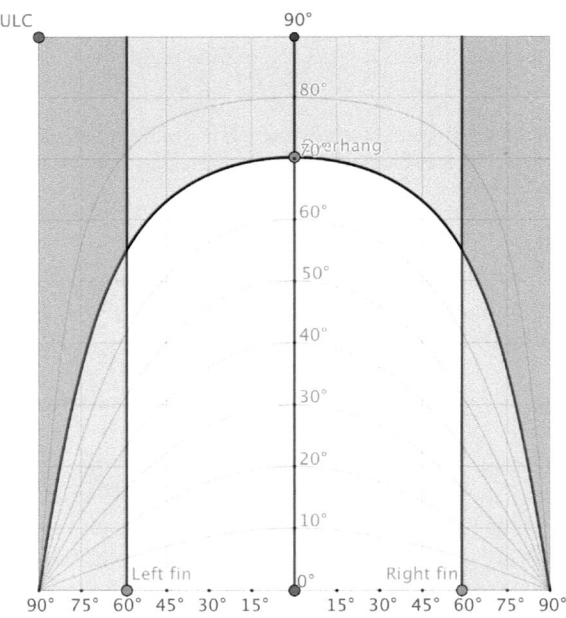

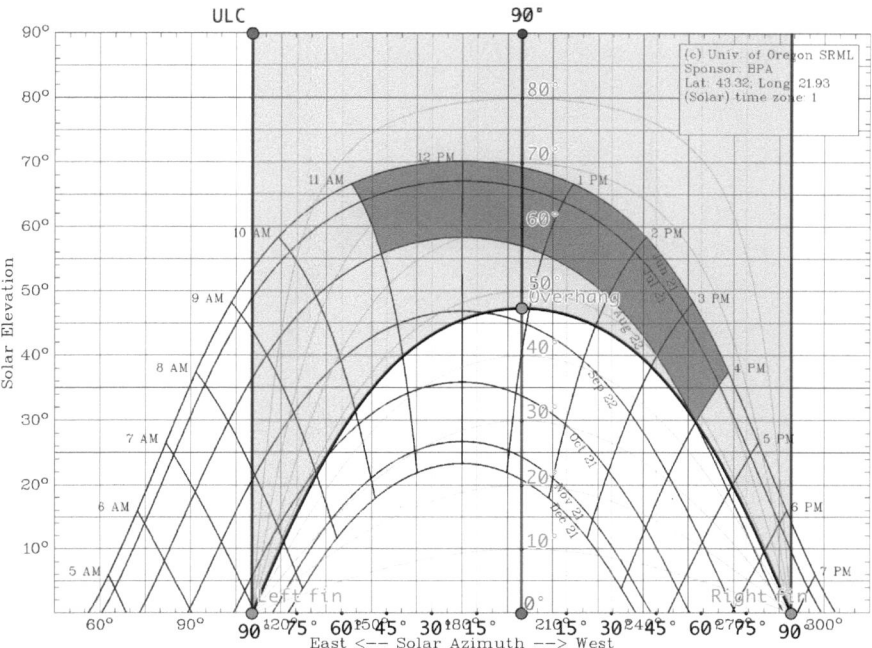

Fig. 2.8 Shading protractor superimposed on a solar path chart in cylindrical projection

azimuth 200° superimposed on the solar path chart from Fig. 2.5. As the two methods are equivalent, we obtain the same result: the point of observation will be shaded during the requested cut-off dates and times if the outer edge of the overhang is visible at an angle of approximately $\alpha = 47°$, while the fins are irrelevant here. Thus, for a window $h = 1.4$ m high and 50% shading, i.e., placing the point of observation in the centre of the window, we obtain that the overhang should have depth $d = 50\% \cdot h / \tan \alpha = 0.65$ m.

A slight advantage of Mazria's cylindrical projection consists in the ease of plotting obstructions visible from the point of observation, as they usually have low latitude at which the chart does not experience much distortion. Hence, it is enough to determine and plot on the chart the horizon's altitude for azimuths in the range at least from 60° to 300° in 15° steps, paying attention to isolated tall objects and deciduous trees, in order to get a fairly precise depiction of surrounding obstructions [19].

Mazria was specifically interested in conterminous US and southern Canada in his book [19], with focus on latitudes from 28 °N to 56 °N. Experiments with shading protractor led him to a general recommendation that in southern latitudes (36 °N), depth of horizontal overhang should be about $\frac{1}{4}$ of the window height, while in northern latitudes (48 °N) it should be about $\frac{1}{2}$ of the window height.

In an effort to facilitate the teaching of designing shading devices and to help understand their impacts, Kensek et al. [14] developed a computer program that

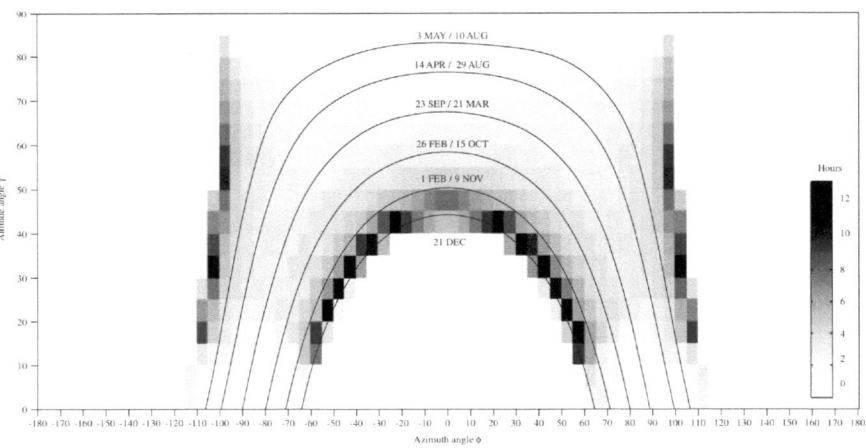

Fig. 2.9 Sky map of annual cumulative probable sunlight duration in Hong Kong [5]

draws the outline of overhang and fins of given dimension ratios upon the solar path chart in Mazria's projection. Bouchlaghem [3] developed another program that implements the method described in [4], which allows the point of observation to be situated in a building's interior, plots the window contours on the solar path chart and calculates times when the point of observation will be illuminated by sunlight. These programs, however, were written more than 20 years ago and do not appear to be available online nowadays.

Solar path charts were improved by Cheung and Chung [5] and Dubois [8] who incorporated additional data into them. Mazria did take cloud cover into consideration by introducing a clearness factor, but for locations in the conterminous US this factor differs only very slightly from one. However, in subtropical and tropical climates with quickly changing cloud cover, the expected duration of sunlight becomes an important property of solar position. This is also supported by Ne'eman et al. [20] who found that the duration of sunlight in the interior, rather than its intensity or the size of sunny patch, correlates best with occupants' satisfaction. Cheung and Chung [5] proposed a method of incorporating information about expected sunlight duration into a solar path chart. They divided the sky hemisphere into 5° × 5° patches and allocated each hourly bright sunshine data measured over a long-term period to the sky patch that contains the position of the Sun in the corresponding hour. Averaging this data enabled them to construct a chart of annual cumulative expected sunlight duration for Hong Kong, as shown in Fig. 2.9. The chart makes it possible to identify smaller sky patches that account for a majority of expected sunlight duration in summer from which the window should be shaded. In the case of Hong Kong, sky patches that appear in two ranges of azimuth angles, −105° to −85° and +85° to 105°, account for more than half of the total summer expected sunlight duration (0° represents South in [5]). Cheung and Chung also discussed optimum configurations of overhangs and side fins for windows of different orientations in an isolated building in Hong Kong with the aim of minimising summer sunlight and maximising the winter sunlight, concluding that it is very difficult to shade the summer sunlight for windows facing east and west due to the two azimuth ranges indicated above.

Dubois [8] incorporated into the solar path chart additional information about the intensity of solar radiation and the solar angle-dependent window properties. Instead of considering just whether the window is lit or shaded at a given time, Dubois was interested in the amount of solar radiation transmitted through the window into the building interior, which determines its solar gain. This value strongly depends on the angle θ of incidence of the solar rays with respect to the window normal: if I_{DN} is the intensity of the direct normal radiation, then the intensity I_θ of direct solar radiation on the window surface is

$$I_\theta = I_{DN} \cos \theta.$$

Further, the window g-value that describes the part of the incident solar radiation, which is absorbed and transmitted by the window as heat into the building, also depends on the incidence angle θ and can be denoted as g_θ. These two values can be combined into a single, cosine weighted solar angle-dependent g-value, or shortly Gcos value, as

$$\mathrm{Gcos}_\theta = g_\theta \cos \theta,$$

so that the solar gain Q_{sol} of the window due to direct solar radiation can be calculated as

$$Q_{sol} = I_{DN} \cdot \mathrm{Gcos}_\theta \cdot A$$

where A is the window area.

The intensity I_{DN} of solar radiation can be included in the solar path chart by using points of different sizes, as shown in Fig. 2.10. Since solar radiation is not symmetrical about the solstices, each point is an average of the two values for symmetrical dates. If Gcos values are further normalised with respect to Gcos_0, then the sets of solar azimuths and altitudes corresponding to specific Gcos_θ represent cones with centre at the point of observation. These cones can be visualised by showing regions with normalised Gcos values in specific ranges ($\mathrm{Gcos}_\theta \in (0.9, 1.0]$, $\mathrm{Gcos}_\theta \in (0.8, 0.9]$, etc.), which are bounded by distorted circles in cylindrical projection, and can be superimposed on the solar path chart. An example of a solar path chart that combines the solar paths with the intensity of direct normal radiation on clear days and the window Gcos values is shown in Fig. 2.10. It has been used by Dubois [8] to design the geometry of an awning for a south- and west-oriented office room in Stockholm, Sweden. The chart helped to identify shading alternatives which were further simulated in Derob-LTH, to precisely identify the optimum awning geometry and compare it with the geometry suggested by the chart. Simulations with Derob-LTH indicated it is most important that the shading mask of a shading device covers regions with higher Gcos values, i.e., those around the window normal.

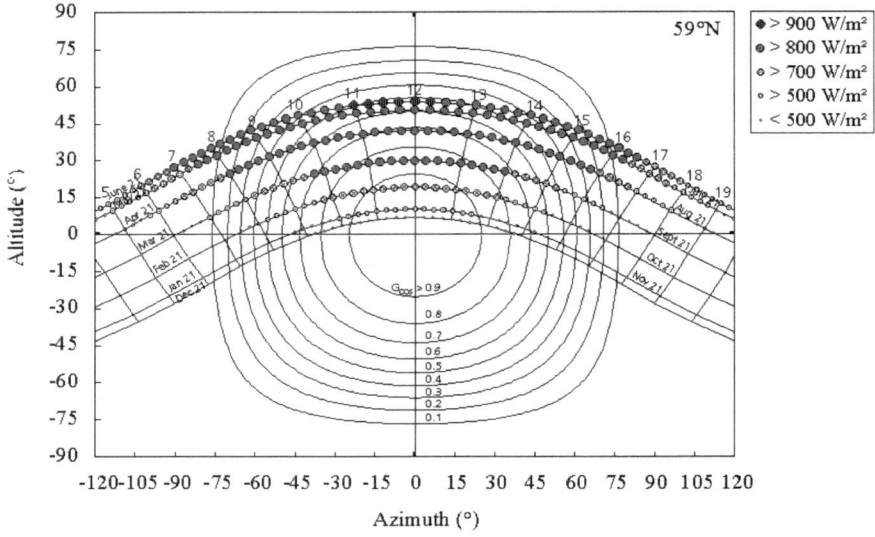

Fig. 2.10 Chart of the normalised Gcos values for a vertical, south-oriented, triple pane, clear glass window superimposed on the solar path chart for latitude 59 °N, showing intensity of direct normal solar radiation (I_{DN}) for clear days in Stockholm [8]

2.3 Backward Tracing of the Solar Rays

Several methods were developed that implement the adage of "full shading in summer and no shading in winter" by raytracing techniques. They do not employ shading masks, but are rather based on tracing the solar rays backward from the window until they intersect a selected shading support structure.

The first such method was described mathematically by Arumí-Noé [2]. It asks for the winter design period, during which the window will be 100% exposed, and the summer design period, during which the window will be 100% shaded, These periods are necessarily symmetric with respect to solstices, and their boundary dates serve as the winter and the summer design day. The first step is to construct a winter solar funnel surface, which guarantees that no point on the funnel surface casts a shadow on the window during the winter design day. This surface is a developable space surface composed of a series of plane surfaces, each of which is associated uniquely with either a window edge or a corner, and critically tilted with respect to the window plane so that the window is never shaded during the winter design day. A shading device is then constructed on the winter solar funnel surface. It is obtained by clipping funnel surface edges with the solar rays on the summer design day and traced backward from each window corner until each funnel surface edge is as short as possible. Consecutive clip points define the edges of the final overhang. The method works for windows having arbitrary convex polygonal shapes and was

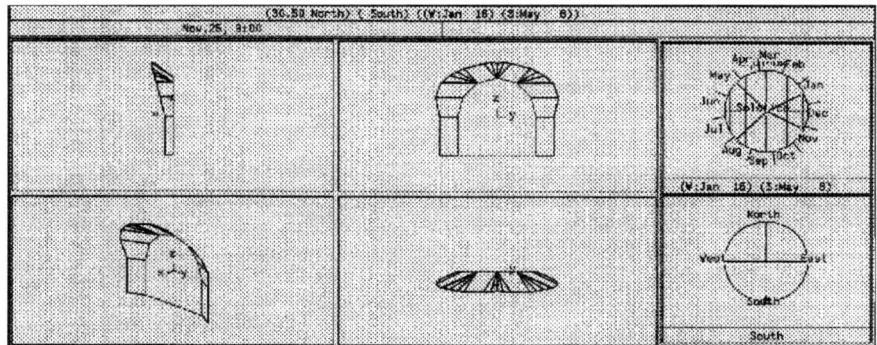

Fig. 2.11 Arumí-Noé's method of clipping winter solar funnel surface applied to an arched window [2]

Fig. 2.12 Illustration of Marsh's method for construction of an overhang shading the window completely between summer cut-off dates [17]

implemented in a computer program, which does not appear to be available online. An example of its application is shown in Fig. 2.11.

Marsh [17] used backward tracing of the solar rays to construct, for an arbitrary shading support surface, the shading device with a minimal area that fully shades the window throughout given summer cut-off dates and times (with cut-off dates symmetrical about the solstices). The resulting shading shape accommodates the curved path of the Sun. Its shape is generated by intersecting shading support surface and the solar rays projected from the vertices forming the windowsill: the sides are determined by the solar rays at each cut-off time throughout the shading period, while the front is determined by the trail of solar rays between the two cut-off times at the cut-off date. For the case of a vertical rectangular window and a horizontal plane as a shading support surface at latitudes further from the equator, the overhang has as its sides appropriate halves of analemmas for the cut-off times, while its front is formed by horizontal segments equal in length to window width and parts of the solar path for the cut-off date traced back from the windowsill vertices (see Fig. 2.12). For more complex cases, rays are generated from points of the shaded surface, back through the shading support surface and surrounding geometry towards the Sun. The result in this case is a point cloud that displays the relative distribution of solar intensity over a shading support surface. An architect then has an option to

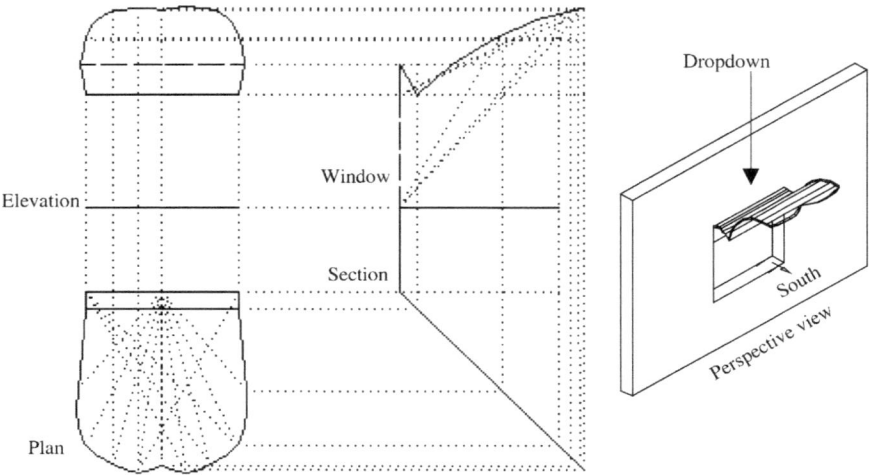

Fig. 2.13 Plan, section, elevation and perspective view of the overhang constructed in [25] for a location of Pilani, India

determine shading shape by setting a threshold on the minimal solar intensity that ought to be shaded. Although less precise, this method can accommodate external obstructions, variable transparency and other concerns that are not taken care of using strict geometric solutions. The method is implemented into Autodesk's Ecotect. It was further elaborated by dividing the shading support structure into an array of cells by both Kaftan and Marsh [13] and Sargent et al. [26] (whose approaches will be described later in Chap. 4).

Ralegaonkar and Gupta [11, 25] presented another, less precisely described overhang design method based on backward tracing of the solar rays. It is based on the selection of two design dates: the winter cut-off date, when the overhang should allow full insolation of the window, and the summer cut-off date, when the overhang should completely shade the window between two given summer cut-off times. The solar rays then appear to be traced back from the upper edge of the window for the winter cut-off date and from the lower edge of the window for the summer cut-off date between cut-off times, with their intersection determining the overhang geometry. Figure 2.13 shows an overhang constructed for a south-oriented window in Pilani, Rajasthan, India, with the winter cut-off date of December 22 and the summer cut-off date of March 23 with the summer cut-off times from 8 am to 4 pm. The overhang support structure has a drain to lead rainwater away from the wall during the monsoon season.

References

1. Arias S (2006) Optimization of the solar control devices in windows for hot climates. In: Proceedings of PLEA 2006: the 23rd conference on passive and low energy architecture, Geneva, Switzerland. Accessed 6–8 Sep 2006
2. Arumí-Noé F (1996) Algorithm for the geometric construction of an optimum shading device. Automat Constr 5:211–217
3. Bouchlaghem NM (1996) A computer model for the design of window shading devices. Build Res Inf 24:104–107
4. Burberry P (1983) Practical thermal design of buildings. Batsford Academic and Educational Ltd, London
5. Cheung HD, Chung TM (2007) Analyzing sunlight duration and optimum shading using a sky map. Build Environ 42:3138–3148
6. Denzer A (2013) The solar house: pioneering sustainable design. Rizzoli International Publications, New York
7. Dias JB, Soares PMM, Graça GGd (2020) The shape of days to come: effects of climate change on low energy buildings. Build Environ 181:107125
8. Dubois MC (2001) A simple chart to design shading devices considering the window solar angle dependent properties. In: Furbo S (ed) Proceedings of EUROSUN 2000: the 3rd ISES Europe solar congres, Copenhagen, Denmark, 2000 Jun 19–22. ISES, Freiburg
9. El-Refaie MF (1987) Performance analysis of external shading devices. Build Environ 22:269–284
10. Etzion Y (1992) An improved solar shading design tool. Build Environ 27:297–303
11. Gupta R, Ralegaonkar RV (2006) New static sunshade design for energy-efficient buildings. J Energ Eng 132:27–36
12. Jaloxa (2022) Sunpath diagrams. https://jaloxa.eu/resources/daylighting/sunpath.shtml. Accessed Feb 2022
13. Kaftan E, Marsh A (2005) Integrating the cellular method for shading design with a thermal simulation. In: Santamouris M (ed) Proceedings of the international conference "Passive and low energy cooling for the built environment", Santorini, Greece, 2005 May 19–21. Heliotopos Conferences, Santorini, pp 965–970
14. Kensek K, Noble D, Schiler M, Setiadarma E (1996) Shading Mask: a teaching tool for sun shading devices. Automat Constr 5:219–231
15. Leatherbarrow D, Wesley R (2014) Performance and style in the work of Olgyay and Olgyay. arq-Archit Res Q 18: 167–176
16. Marsh A (2022) 2D Sun-Path. https://andrewmarsh.com/apps/releases/sunpath2d.html. Accessed Feb 2022
17. Marsh A (2003) Computer-optimised shading design. In: Proceedings of building simulation 2003: the 8th international IBPSA conference, Eindhoven, Netherlands, 2003 Aug 11–14. IBPSA, pp 831–837
18. Matusiak M (2022) Solarbeam. https://solarbeam.sourceforge.net. Accessed Feb 2022
19. Mazria E (1979) The passive solar energy book: a complete guide to passive solar home, greenhouse and building design. Rodale Press, Emmaus
20. Ne'eman E, Light W, Hopkinson RG (1976) Recommendations for the admission and control of sunlight in buildings. Build Environ 11:91–101
21. Oh JKW, Haberl JS (1997) New educational software for teaching the sunpath diagram and shading mask protractor. In: Proceedings of building simulation 1997: the 5th international IBPSA conference, Prague, Czech Republic, 1997 Sep 8–10. IBPSA, cód P203
22. Olgyay V (1957) Solar control and shading devices. Princeton University Press, Princeton
23. Olgyay V (1963) Design with climate: bioclimatic approach to architectural regionalism. Princeton University Press, Princeton
24. Prinsloo GJ, Dobson RT (2015) Solar tracking. SolarBooks, Stellenbosch
25. Ralegaonkar RV, Gupta R (2005) Design development of a static sunshade using small scale modeling technique. Renew Energ 30:867–880

26. Sargent JA, Niemasz J, Reinhart CF (2011) Shaderade: combining Rhinoceros and Energyplus for the design of static exterior shading devices. In: Proceedings of building simulation 2011: the 12th international IBPSA conference, Sydney, Australia, 2011 Nov 14–16. IBPSA, pp 310–317

27. Stevanović S (2022) Interactive shading protractor in GeoGebra for Olgyays' stereographic projection. Zenodo https://doi.org/10.5281/zenodo.2576560. Accessed Feb

28. Stevanović S (2022) Interactive shading protractor in GeoGebra for Mazria's cylindrical projection. Zenodo. https://doi.org/10.5281/zenodo.2579779. Accessed Feb

29. Subhashini S, Thirumaran K (2018) A passive design solution to enhance thermal comfort in an educational building in the warm humid climatic zone of Madurai. J Build Eng 18:395–407

30. Szokolay SV (2022) Solar geometry. Brisbane: PLEA passive and low energy architecture international and department of architecture, The University of Queensland. https://www.plea-arch.org/wp-content/uploads/PLEA-NOTE-1-SOLAR-GEOMETRY.pdf. Accessed Feb 2022

31. Tahbaz M (2012) Primary stage of solar energy use in architecture-Shadow control. J Cent South Univ 19:755–763

32. University of Oregon Solar Radiation Monitoring Laboratory. Polar sun path chart program. https://solardat.uoregon.edu/PolarSunChartProgram.html. Accessed Feb 2022

33. University of Oregon Solar Radiation Monitoring Laboratory. Sun path chart program. https://solardat.uoregon.edu/SunChartProgram.html. Accessed Feb 2022

Chapter 3
Current Overhang Research Methodology

Abstract In this chapter, we review the methodology used in current overhang research. The first section discusses simulation methods and algorithms, both those developed specifically for overhangs and those developed for general shading calculations in simulation engines. The second section is concerned with the ways of quantifying overhang effectiveness in the research literature, from the viewpoints of both thermal and visual performance. Genetic algorithms are the most used optimisation approach in overhang design, as well as in general building energy optimisation. The third section, in addition to describing the work and practical setup of genetic algorithms, reviews the overhang studies employing them.

Keywords Shading masks · Shading algorithms · Objective function · Thermal performance · Visual performance · Multi-objective optimisation methods · Pareto front · Genetic algorithms

3.1 Simulation Methods

Here, we review simulation methods and algorithms for calculation of shading factors provided by overhangs. The first subsection reviews mainly older methods that were developed particularly for overhangs, while the second subsection reviews general methods—polygon clipping and pixel counting—that are used for shading calculations in modern simulation engines. The third subsection discusses speeding up of computations by grouping solar positions into small sky patches, while the last subsection, on the other hand, considers the influence of solar positions and uncertainties of input parameters on the precision of shading calculations.

3.1.1 Methods Developed Particularly for Overhangs

Jones [26] developed one of the oldest theoretical methods in 1980 to calculate monthly average daily insolation on a window of arbitrary azimuth shaded by an overhang, by extending previous methods of Liu and Jordan [28], for calculating insolation on surfaces tilted towards the equator, and of Klein [34], for calculating insolation on surfaces whose normals have components towards the equator. Besides direct radiation, Jones considered diffuse and ground reflected radiation, assuming both to be isotropic, but neglected reflections by the overhang and surrounding walls. He also assumed the overhang to extend laterally far past the sides of the window, i.e., that it has infinite length, as the shadow cast by the overhang on the window then has a simple shape: a rectangle of width equal to that of the window, fully determined just by its height. Yanda and Jones [108] extended Jones' method to overhangs of finite lateral width, but only for windows facing the equator.

Sharp [83] presented another theoretical method to calculate monthly average daily insolation on a window shaded by an overhang. In addition to direct, diffuse and ground reflected radiation, he also considered radiation reflected from the underside of the overhang onto the window. This approach allowed the window to have arbitrary azimuth and arbitrary tilt, but assumed the overhang to be of infinite length.

A bit earlier, Utzinger and Klein [96] developed in 1979 a graphical method to estimate monthly average daily radiation on vertical windows shaded by overhangs. Their approach allowed the window to have arbitrary azimuth, included also radiation reflected from the overhang onto the window, and was applicable to overhangs of finite lateral width. Rahman et al. [64] drew attention to the fact that the method of Utzinger and Klein [96] does not include atmospheric transmittance into consideration, which, as it turns out in their discussion, has particularly significant influence for non-southern orientations at lower latitudes. Jones [26] dealt with this issue by using a constant value of 0.8 as approximation for the ratio of daily insolation on a horizontal plane on the ground with that on an extraterrestrial horizontal plane.

Raeissi and Taheri [63] in 1998 extended the method of Yanda and Jones [108] to estimate instantaneous irradiance, not just monthly average or clear day insolation, on windows of arbitrary azimuth and with overhangs of finite lateral width. They based their method on empirical equations of Daneshyar [12] for instantaneous direct and diffuse solar radiation on horizontal flat surfaces in Iran, although the method can be adapted to use analogous equations for other locations. They also consider radiation reflected from the underside of the overhang onto the window. A method was implemented in a computer program for the calculation of hourly cooling load requirements for a building and validated by comparison with field data from an actual house in Iran.

Włodarczyk and Nowak [107] developed in 2009 an analytical method, based on Perez's assumption about anisotropic solar radiation onto inclined plane [56], to calculate solar gain reduction from the presence of overhang and compare its results to both measured data and results of EnergyPlus simulations, obtaining a good agreement after taking into account slightly different climatic data used for EnergyPlus simulations. Results are presented in the form of nomograms, obtained by least squares fitting, that show solar gain reductions with respect to ratios P/H and G/H, where P is overhang depth, G is vertical distance between overhang and window upper edge, while H is window height. The nomograms show, for example, that if P, G and H are selected so to reduce summer solar gains by 50%, then this also reduces winter solar gains by 15–25% in Polish climatic conditions.

Sameti and Jokar [79] in 2016 developed a mathematical model to simulate solar thermal energy transfer through windows with overhangs of finite width, by working out in detail feasible shapes of a shadow cast by overhang. In their model, they consider direct radiation, isotropic diffuse and ground reflected components and anisotropic circumsolar diffuse component and horizon diffuse component from the Perez sky model [56], while disregarding part of the radiation reflected from the underside of the overhang onto the window.

While previously mentioned methods were interested in either cooling loads or solar thermal gains, Rao and Tzempelikos [66] studied the daylighting performance of exterior overhangs. Overhang is treated as a form of a light shelf in this case: it is placed in a horizontal plane that splits a window, reflecting direct solar radiation incident on it to the upper part and shading the lower part of the window. The method assumes an infinite overhang, i.e., that the unshaded portion of the window is a rectangle of the same length as the window. The sky model of Perez et al. [56], which considers the sky as an anisotropic source of diffuse light, is used to calculate the hourly sky diffuse illuminance on the façade, while a multiple bounce radiosity method, using fundamental geometrical equations of Siegel and Howell [85], serves to predict work plane illuminance. The method computes annual daylight autonomy and fractions of shaded windows and sunlit floor areas as functions of overhang geometry and reflectance throughout the year.

3.1.2 Methods for General Shading Calculations

While the geometry of overhangs and side fins is simple enough to allow the development of specialised methods mentioned in the previous section, it is nowadays common to perform building performance simulations within general simulation engines such as EnergyPlus, TRNSYS, IDA ICE, …which require methods capable of performing shading calculations for building elements of more complex geometry. Most simulation engines represent building geometry with polygons and they usually resort to polygon clipping methods in which shading factors are obtained by projections along the direction of the solar rays. In these methods, a shadow is projected from the vertices of each shadowing polygon to the plane of the polygon receiving shadow. However, if any vertex of the shadowing polygon is on the opposite side of the plane of the shadow receiving polygon from the Sun, then a false shadow would be cast, so the shadowing polygon must be clipped off by this plane before projection [6].

EnergyPlus, the most used simulation engine in current shading studies [33], offers the choice of two such clipping algorithms: the Sutherland–Hodgman and the convex Weiler–Atherton. The Sutherland–Hodgman algorithm [93] is a simpler and faster algorithm, and can also be used when exterior wall surfaces receiving shadows are not convex. The Weiler–Atherton algorithm [103] is general enough to process almost all kinds of non-self-intersecting polygons, but its implementation in EnergyPlus nevertheless requires both casting and receiving surfaces to be convex. In addition to the choice of polygon clipping algorithm, two further parameters are also important for the proper calculation of shading effects when performing shading simulations with EnergyPlus [91]:

- the calculation method field of the *ShadowCalculation* object has to be set to *TimeStepFrequency* to perform solar path, shadowing and diffuse sky modelling calculations at each time step.
- the solar distribution field of the *Building* object has to be set to *FullInteriorAndExteriorWithReflections* to calculate reflections of direct and diffuse solar radiation from exterior surfaces, compute shadow patterns by the window shading and calculate amounts of transmitted direct radiation falling on each internal surface by projecting the solar rays through the window.

Källblad [30] described a polygon clipping method implemented in Derob-LTH (short form *Dynamic Energy Response Of Buildings-Lund Institute of Technology*). It uses a recursive formula to determine the non-shaded area of a surface lit by direct radiation with an arbitrary number of shadows cast by other surfaces, including radiation reflected from surfaces, but under the restriction that all surfaces are planar and convex. Diffuse radiation is assumed to be isotropic and its shading is estimated by calculating view factors from the surface to the sky.

Maestre et al. [39] presented a method that projects every polygon to the unique plane orthogonal to the solar rays first, after which unions of intersections between shading and receiving polygons are obtained by a variant of Vatti's clipping algorithm [100] that works with arbitrarily shaped polygons. Since polygons are projected only once, instead for each pair of shading and receiving polygon, this method may offer potentially improved computational times.

More accurate shading and daylighting results may be obtained by using raytracing, which traces reflections of representative rays of light in a digitally modelled scene to extrapolate the global distribution of light available in the scene. This technique is predominantly used in daylight simulation and computation of illuminance-based metrics. However, the higher precision of raytracing is offset by substantially longer computation times. Radiance and DAYSIM, which are based on Radiance, are currently the most utilised tools for daylight simulation [82]. Comparing the results of EnergyPlus and Radiance in the evaluation of useful daylight illuminance, Ramos and Ghisi [65] observed the limitations of EnergyPlus associated with light decay in deep rooms. Using the CIE 171:2006 protocol for daylighting simulation, Acosta et al. [1] found that EnergyPlus has an average margin of error of 14%, which can go up to 40% at certain points. Queiroz et al. [61] confirmed that EnergyPlus has an average error of 14% when compared to the results of Radiance, with a standard deviation of 6%, and that it can significantly diverge for complex geometries. However, they also observed that EnergyPlus classified solutions similar to Radiance, with the linear correlation coefficient $R^2 = 0.90$ between all data for the useful daylight illuminance of a space with no overhang and with two different overhangs. Due to this and the errors within a 20% margin, which are deemed acceptable for tools used in early design stages by Nielssen [54] and Hviid et al. [23], approximate calculations of EnergyPlus can still be considered satisfactory for simulations of low complexity façade designs.

Another general method for the calculation of direct radiation shading is pixel counting. It uses computer graphics methods to render the building scene in orthogonal projection from the viewpoint of the Sun, once with shading devices and obstructions and once without them. Pixels in the shade get the colour of shading elements, different from those exposed to direct radiation, so counting pixels by colour for each surface in the two renderings determines the sunlit fraction of that surface. Unlike polygon clipping methods, this approach works for any building geometry without any restriction. It was proposed and developed by Yezioro and Shaviv [84, 110] and Niewienda and Heidt [55]. While the precision of obtained results is determined by the resolution used for renderings, Jones et al. [27] showed that the relative error is within 1% of analytical values when used to calculate the sunlit fraction of building façades. Rocha et al. [74] have further compared and validated both polygon clipping methods implemented in EnergyPlus and pixel counting methods implemented in Shading II plug-in for SketchUp [110] and Domus simulation engine [48] with data measured from small-scale mock-up models. Their results show that for a prototype with a simple, comb-like shading geometry, all methods give good agreement with measured data. However, for a prototype with a hollowed shading screen, polygon clipping methods were not able to produce meaningful results due to their geometric limitations, while pixel counting methods still gave an accurate assessment of sunlit surface fraction.

3.1.3 Grouping of Solar Positions in Shading Calculations

One way to speed up shading calculations is to group sufficiently similar solar positions that produce only slightly differing fractions of sunlit areas of building surfaces and then perform just a single shading calculation for the representative solar position of each group.

Mardaljević [45] described an image-based method for quantifying the effectiveness of shading devices in blocking direct solar radiation. The technique is based on rendering the surface in question with the Radiance lighting simulation system [102], where a green "wash" of ambient light is used to provide a subdued background against which the bright patches of solar penetration are clearly seen, so that pixel counting can then be used to estimate the potential for solar gain and the associated cooling load. To reduce the number of renderings, solar altitude and azimuth angles are distributed into sky patches using a grid with specified angular resolution (8° in this case). The Sun positions that occur in each sky patch are counted and the average Sun position is evaluated, after which an irradiance rendering showing the surface is generated for each patch with the unit radiance Sun placed at the average Sun position. The total annual irradiation image can then be synthesised from the collection of these renderings, after taking into account the average radiation intensity within each patch. In [45], this approach was used for a complex roof shading system consisting of 3600 parametrically generated fins.

Marsh [46] describes the use of shading masks, as implemented in Autodesk Ecotect, in performance gains, particularly in the cases when they can be kept between different runs of building performance simulation. The sky is similarly divided into equal-angle patches for simpler processing of corresponding data structures. A shading mask is then simply a mechanism for recording sky patches that are visible from a particular point, while partial shading of a surface is determined by sampling uniformly distributed points from the surface and averaging the results into a single shading mask. The paper also explains how the information about direct and anisotropic diffuse solar components can be combined within the same shading mask.

3.1.4 Sensitivity Analysis of Shading Calculations

A common practice to speed up building performance simulation is to determine shading factors for an evenly distributed subset of days during the simulation period only (which should not be confused with the grouping of similar solar positions into sky patches). Maestre et al. [38] quantified relative errors made when shading factors were calculated this way for a number of overhang and side fin configurations, window orientations and several locations. EnergyPlus was used with shading calculations performed every 20 days, and results were compared with an analytical solution from [3]. While deviation in annual energy consumption was just 0.37%, at the monthly level it reached a relative error of 7.5%, while it was even higher on daily (19%) and hourly (37%) levels. Moreover, about 25% of all cases, including a frequent configuration of the overhang with depth equal to half of the window height, have an annual relative error larger than or equal to 5%. Maximum relative errors occur during the spring and the autumn months, due to a higher variation of solar positions in these months.

Rocha et al. [73] extend the work of Maestre et al. [38] by performing ANOVA-based sensitivity analysis of the sunlit area of the building exterior surface with shading devices for six selected input parameters. The shading device used is a comb-like overhang, placed on an exterior wall without windows. The Sutherland–Hodgman algorithm is employed as a shading algorithm in EnergyPlus, and solar calculations were performed for locations of Miami and New York for four specific days: March 21, June 21, September 21 and December 21. Results show that more than 20% of all simulation time steps have uncertainties larger than or equal to 5% on the solar fraction calculation at 95% confidence level. Actually, such uncertainties appear for about 40% of the time steps in Miami compared to only 6% of time steps in New York, with maximum values of uncertainties occurring in warmer months, which is explained by higher solar paths in Miami where overhang more strongly influences the sunlit area. Three input parameters, latitude, building orientation and width of overhang teeth, have a higher influence on output uncertainties, where a change of only 5° in building orientation may lead to significant changes in the fraction of the sunlit area.

3.2 What to Measure?

The thermal performance of overhangs is reflected in its influence on solar gains through the associated window, and most studies use an ideal loads air system in simulating their building models to quantify this influence through changes in heating and cooling loads. Visual performance is reflected in the influence of overhangs on the availability of daylight in interior spaces, which is often measured through changes in artificial lighting loads only. In their in-depth review, Yu et al. [111] show that most research studies aim to reach a balance between thermal and daylighting performance by considering a combination of performance indicators that describe thermal energy, daylighting energy and daylighting comfort. We discuss here thermal and visual performance indicators that are used to quantify overhang effectiveness in the research literature.

3.2.1 Thermal Performance Indicators

Annual heating and cooling loads obtained by simulating ideal loads air system are almost ubiquitous in overhang design studies, as only two studies use somewhat different thermal performance indicators.

As a matter of fact, Loonen and Hensen [36] base their sensitivity study on instantaneous heating and cooling loads calculated at each simulation time step. Sensitivity indices for depths of overhang and two side fins, as input variables, are functions of time step showing a high degree of short-term fluctuations. Loonen and Hensen pinpoint the importance of using moving averages to smooth out these fluctuations [98], which can effectively visualise time-varying relations between shading parameters and heating and cooling loads and reveal design parameters that are influential at particular times of the year. Namely, when sensitivity indices with standardised regression coefficients are calculated at an annual level, they show that overhang depth has the most significant impact on heating and cooling loads, while the impacts of side fins are much smaller and almost equal. Their negative sign shows that increasing shading element depths will decrease annual heating and cooling loads, but cannot distinguish between their impact in the summer and winter months. On the other hand, calculating 24-h moving average of sensitivity indices smooths out (sub)hourly variations and focuses attention on changes on a daily level. This reveals that the eastern fin has a significantly smaller impact than the western fin during most part of the year, which is expected as the window has a south-east orientation, so the eastern fin blocks only early morning solar radiation in summer, while western fin blocks solar radiation later in the day, when its influence on cooling loads is much higher. Further, calculating the 10-day moving average of sensitivity indices reveals information on their longer term trends. It easily shows that at a latitude of 40 °N with higher summer solar altitude angles, the overhang depth is almost the only sensitive variable in the summer months, as side fins block only a small amount of solar radiation then. It also demonstrates that shading has a significant negative effect in winter, which is counterbalanced by its positive effects in summer. A sudden change in signs of sensitivity indices for the overhang and eastern fin depths observed around the end of November and the end of January suggests that it may be worthwhile to design shading as an add-on façade element and at the same time proposes adequate transition moments between winter and summer modes. However, Loonen and Hensen also caution about using this type of analysis in cases when instantaneous effects of shading may be spread over several time steps, such as in buildings with large thermal mass, which may cause too much interference to yield useful results.

Sghiouri et al. [80] apply single-objective genetic optimisation with building energy simulation to find optimal depths of overhangs in different thermal zones of a building that minimise discomfort degree hours. The adaptive comfort temperature model is used according to the standard EN15251, as the building is naturally ventilated and located in a hot climate, where occupants can adapt to the temperature felt to improve their comfort. In the summer season, the comfort temperature T_c is calculated according to the weighted mean of the previous 7-day external air temperatures and discomfort degree hours are calculated with respect to the range $[T_c - 3°, T_c + 3°]$, while in the remaining seasons temperatures out of the range $[20°, 26°]$ are considered as uncomfortable. This model leads to smaller cooling needs in summer resulting in smaller overhangs obtained as optimal solutions. However, this study is limited in the sense that windows of a naturally ventilated building are assumed to be permanently closed, assuming only that there is a fixed infiltration rate of 0.6 ach.

3.2.2 Daylighting Performance Indicators

Many studies have confirmed positive influences of daylight in interior spaces [52], and access to daylight nowadays finds a well-deserved place in sustainable building assessment methods [88]. Moreover, Turan et al. [95] were able to estimate, by inputting the daylight simulation results into a hedonic pricing model, that high daylight imposes 5–6% financial premium in office rent prices in Manhattan.

However, when it comes to measuring the influence of shading on daylight availability, there is still no consensus about the "standard" performance indicator [82]. Unlike heating and cooling loads that are described by a single number for the whole zone, daylight illuminances differ from point to point in the zone, typically diminishing rapidly with increased distance from a window. While artificial lighting load does describe the overall contribution of daylight in satisfying illuminance requirements of a zone, it cannot distinguish whether and for how long parts of the zone have insufficiently lit work planes or unacceptably high glare. The following indicators are some of the proposed ways to quantify daylight availability in overhang shading studies.

Daylight factor DF [67] is an older indicator, introduced in the UK in the first half of the twentieth century [101], that represents a ratio between illuminance measured at a reference point inside and illuminance under standard CIE overcast sky outside. While it may be suitable for temperate climates at higher latitudes with mostly cloudy skies, it is neither based on realistic climate data used for estimating solar gains in building performance simulations nor adequate for climates with more annual sunshine hours, so other climate-based indicators were proposed instead.

Daylight autonomy DA [52, 71] represents the percentage of working hours during the year when the minimum work plane illuminance requirement is met by daylight alone under realistic, time-varying sky and sun conditions. It depends on the illuminance requirements of the user, which range typically from 300 to 500 lux [13, 66]. On the other hand, spatial daylight autonomy sDA [24] represents the percentage of the floor area of a given room at which daylight exceeds a specified illuminance (e.g., 300 lux) for a specified percentage of occupied hours (e.g., 50%).

Useful daylight illuminance UDI [51, 52] improves daylight autonomy by excluding from consideration illuminances that are too bright. Specifically, it measures the percentage of working hours during the year when daylight illuminance observed on the work plane is in a predefined range of useful illuminances based on human factors. Since people tend to tolerate much lower illuminance levels of daylight than artificial light [4], Nabil and Mardaljević [51, 52] set the lower limit of useful range to 100 lux. The upper limit was set to 2000 lux, taking into account results of earlier experiments [76, 77] showing that a properly designed visual environment was still considered reasonably comfortable when the work plane illuminance was below 1800 lux. Shafavi et al. [81] surveyed the visual comfort of students in 20 architectural studios, aiming to correlate it with the various illuminance-based metrics. No solid correlation was observed both because these summarising metrics cannot describe daylight distribution in space and because visual comfort can be achieved within a range of illuminances based on occupants' activity, due to eye adaptation. On the other hand, their study did show that the useful daylight illuminance $UDI_{300-3000lx,50\%}$ is relatively consistent in predicting the students' perception of daylight sufficiency. Related to overhangs, Esquivias et al. [17] show in a case study located in Seville, Spain, that overhangs fitted on the south-facing façade are most effective in reducing excessive illuminance in interior spaces. While overhangs can lead to a major reduction in daylight autonomy and may increase artificial lighting demand, the UDI results indicate that most of the work plane surfaces still remain in the range 100–500 lux.

The useful illuminance range may also be defined according to research needs: for example, David et al. [13] use 300 lux as the minimum and 8000 lux as the maximum acceptable illuminance. When calculating UDI for cellular office spaces, illuminance across the whole core of the work plane has to be in useful range, while for large, open places UDI values are calculated for each point independently. Selecting two threshold values to define useful illuminances further enables one to also quantify occurrences when daylight illuminance falls short of the useful range or exceeds the useful range, leading to possible glare or overheating problems. To quantify these issues, David et al. [13] have further defined the sun patch index SP as the percentage of the working area where the level of illuminance is higher than 8000 lux, integrated over all working hours. On the other hand, glare is usually quantified through more common glare metrics such as daylight glare index DGI [22] and daylight glare probability DGP [106], whose limitations in brightly lit spaces were recently discussed in [47]. Among overhang studies, Yener [109] had used DGI, and calculated analytically according to [72], in the study of a room with a horizontal overhang. Yet another possible predictor of glare is annual sunlight exposure ASE [24] that represents the percentage of the work plane at which direct solar radiation exceeds a specified illuminance level (e.g., 1000 lux) for more than a specified number of hours (e.g., 250 h) during occupied hours (with all operable shading devices retracted). However, recent measurements in an office building in San Francisco [99] indicate that ASE may overpredict the actual occurrence of glare.

An interesting alternative for measuring daylight availability has been recently proposed by Manzan and Padovan [43]. Namely, in the study model they implement an automated blind control system, with blinds fully lowered as soon as direct solar irradiance is above $50\,W/m^2$ on sensors and fully open when it is below this limit, which is proposed as an ideal blind control system that maximises daylighting in [69]. Two sensors are placed at work plane height at distances 1 and 2 m from the window, and the percentage of time when blinds are lowered is treated as the daylight performance indicator.

An attempt to quantify uniformity of daylight availability with a single number may be found in [35], where the two ratios of the maximum and the minimum illuminance, and of the average and the minimum illuminance, were used as the indicators of illuminance uniformity in an office space.

3.3 Optimisation Approaches

While shading optimisation problems are sometimes defined as single-objective, using primary energy demand to combine thermal and visual performance indicators of the overhang, most such studies use multi-objective optimisation, studying the corresponding Pareto front in the search for optimal overhang designs. Genetic algorithms have become ubiquitous in building energy optimisation. In the second subsection, in addition to describing their work and practical setup, we review overhang studies employing genetic algorithms, including recently used surrogate models for faster prediction of the Pareto front.

3.3.1 Pareto Front

As we have seen in the previous chapter, there are quite a few ways in which various aspects of overhang performance can be represented. In certain cases, different performance indicators can be combined into a single value, usually as a linear combination with constant coefficients. For example, we may have a building model a, say a cellular office with a single window and an overhang above it, that employs an ideal loads air system for which building performance simulation returns the heating load $H(a)$ as the annual energy for district heating, the cooling load $C(a)$ as the annual energy for district cooling, and the lighting load $L(a)$ as the annual amount of electricity used for artificial lighting. These three loads can be combined into a single value by determining the primary energy needed for heating, cooling and lighting. Assuming that district heating, district cooling and artificial lighting have efficiencies e_H, e_C and e_L, respectively, and source energy factors f_H, f_C and f_L, respectively, the primary energy needed would be equal to

$$E(a) = \frac{f_H}{e_H} H(a) + \frac{f_C}{e_C} C(a) + \frac{f_L}{e_L} L(a),$$

and our task would become to design an overhang that minimises $E(a)$. Variants of this approach have been used in [41–44, 91, 92].

However, as shading devices have to reconcile opposite objectives—preventing excessive solar gains in warm periods without sacrificing beneficial solar gains in cold periods and visual comfort in interior spaces—researchers tend to define optimality in terms of multiple objectives that describe the thermal and visual performance of the overhang. In such cases, Pareto front describes the set of potentially optimal solutions. As described by Stevanović et al. [92], a solution a of an optimisation problem P with objective functions $C_1, C_2, \ldots, C_k$, whose values should be minimised, is a Pareto solution if the values $C_1(a), C_2(a), \ldots, C_k(a)$ are not simultaneously dominated by any other solution, i.e., if there does not exist any other solution b of P such that all inequalities

$$C_1(b) \leq C_1(a), \quad C_2(b) \leq C_2(a), \quad \ldots, \quad C_k(b) \leq C_k(a)$$

hold simultaneously. The Pareto front, which is a set of all Pareto solutions, is usually found by the simple cull algorithm [60]. It starts with an empty set PF and proceeds iteratively through all solutions a that are being found by the optimisation process. Each such solution a is compared to each solution b already in PF: if b dominates a, then a is discarded, while if a dominates b, then b cannot be a Pareto solution, so b is removed from PF. Otherwise, if a is incomparable to all solutions in PF, then a is added to PF as well. At the end of this iterative process, which takes quadratic time in the worst case, the set PF will contain the Pareto front.

Probably, the earliest overhang shading study that employed the Pareto front is [29]. In it, Kabre described her program Winshade for optimising the shading device according to two objectives: the *shading efficiency*, defined as the ratio between the radiation intercepted by the shaded window and the radiation incident on the unshaded window during overheated period, and the *heating efficiency*, defined as the ratio between the radiation entering through the shaded window and the radiation incident on the window during underheated period. The program considered the period of overheating as a variable as well: if climate data were available, it could use either dry bulb temperatures or tropical summer index [10] to help with defining feasible overheating periods, but otherwise it used the start and end of the overheating period as variables to help find an acceptable compromise between shading and heating efficiencies. The reason for such variable choice of overheating period lies in the way in which a potentially optimal shading is constructed: instead of directly constructing overhang and side fins, the program sets up a two-dimensional grid of points over the glazing with user-defined spacing, and then determines the length of perpendicular pins placed at these points such that each pin casts a shadow long enough to reach the periphery of the window at any hour of the overheated period when the sun shines on the window. By varying the start and end of the overheating periods, the program is then able to exhaustively generate the Pareto front and rank Pareto solutions by a weighted sum of shading and heating efficiencies.

A Pareto front may be easily visualised when the optimisation problem has two objective functions $C_1(a)$ and $C_2(a)$, whose values are then depicted on the x and y axes, respectively. If the optimisation process generates tens of thousands of feasible solutions, as is customary in modern building energy performance studies, the Pareto front may turn out to contain hundreds of Pareto solutions. However, if $C_1(a)$ and $C_2(a)$ may be combined into a single objective, for example by translating them into primary energy, by a linear combination

$$T(a) = pC_1(a) + qC_2(a)$$

for some coefficients p and q, then a simple geometric algorithm may lead to a substantially smaller subset of Pareto solutions with optimal values of $T(a)$ for varying values of p and q. Namely, as already observed in [89], the set of solutions with the same value of $T(a)$ represents a line with the slope $-\frac{1}{r}$ for the ratio $r = q/p$. The Pareto solution a with the minimum value of $T(a)$ then belongs to the leftmost of these lines, so that all other solutions are found to the right of this line, which means that such a solution necessarily belongs to the convex hull of the Pareto front. The convex hull is defined as the smallest convex polygon that contains all Pareto solutions. It can be obtained by Fortune's variant of Graham's scan method [19], and it usually contains a much smaller number of Pareto solutions (dozens instead of hundreds in current optimisation studies), which facilitates their analysis and discussion. Moreover, each Pareto solution that is a vertex of the convex hull belongs to two of its sides, say k and l, and it represents a Pareto solution with the optimal value of $T(a)$ for all choices of p and q for which the slope $-\frac{1}{r}$ falls between the slopes of sides k and l. Such an approach was, for example, used in a study by Stevanović et al. [92], where the number of Pareto solutions for heating and cooling loads of an apartment room with a NURBS-lined overhang at different locations

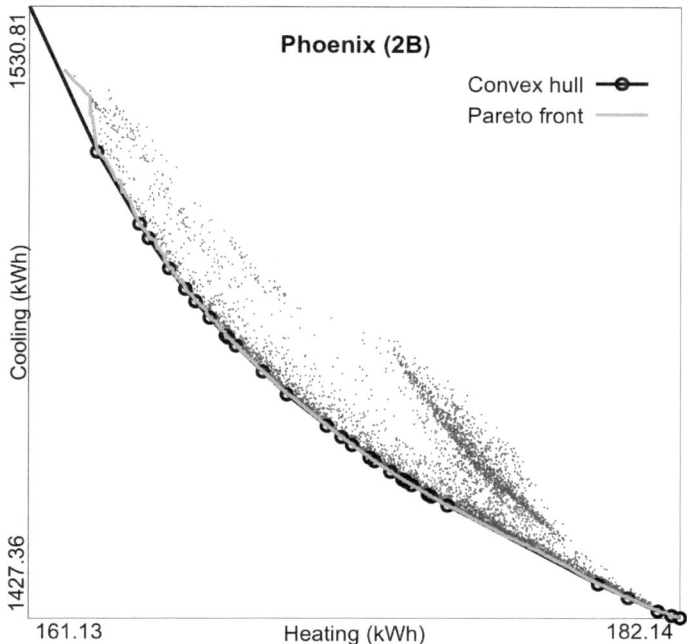

Fig. 3.1 An example from [92] showing Pareto front and its convex hull for a set of solutions characterised by their heating and cooling loads. The Pareto front contains a total of 884 solutions, while its convex hull is determined by only 30 of them

ranged from 783 to 1246, while the number of Pareto solutions in the convex hull ranged from 5 to 41 only. Figure 3.1 shows an example of a Pareto front for one of the locations considered in [92], with convex hull solutions shown with larger dots.

Rocha et al. [75] recall the physical programming [50] as a way to rank Pareto solutions according to predefined user preferences, when the objective functions $C_1, C_2, \ldots, C_k$ cannot be naturally summed up using appropriate weights. In this case, for each objective function C_i the user specifies boundary values which determine the ranges in which its values over all considered building models a may fall: HD (a highly desirable range), D (an acceptable range that is desirable), T (an acceptable range that is tolerable), U (a range that is undesirable) and HU (a range that is highly undesirable). The piecewise linear functions $P_1(C_1(a)), P_2(C_2(a)), \ldots, P_k(C_k(a))$ may then be constructed so that P_i translates the value $C_i(a)$ into the range $[0, 1]$ for $i = 1, \ldots, k$ in such a way that the HU, U, T, D and HD ranges are, respectively, mapped to the intervals $[0, \alpha], [\alpha, \beta], [\beta, \gamma], [\gamma, \delta]$ and $[\delta, 1]$ for selected values of α, β, γ and δ. Then one can rank Pareto solutions by the sum $C_{pp}(a) = \sum_{i=1}^{k} P_i(C_i(a))$, which combines the initial objectives $C_1(a), C_2(a), \ldots, C_k(a)$ according to the predefined user preferences, or alternatively, one can use a single-objective optimisation to find building models with extremal values of C_{pp}. Further improvements to the concept of physical programming were suggested by Tappeta et al. [94].

3.3.2 Genetic Algorithms

As Loonen and Hensen observed in [36], larger shading elements are not by definition better, and careful dimensioning is essential to achieve high performance. To accommodate selected thermal and daylighting objectives, newer shading studies usually turn to optimisation algorithms. The *search space* is defined first by listing feasible values for the usually independent parameters that describe a particular solution. These search spaces are usually prohibitively large to allow their exhaustive exploration, so that an optimisation method is employed to produce (nearly) optimal solutions through a number of building performance simulations. Parts of this section are based on the author's recent papers [91, 92].

Coupling of building energy simulation tools with optimisation methods had become mainstream in the study of energy and buildings after Caldas and Norford [9] used it prominently to facilitate performance-based façade design. A number of reviews on this topic are available: Machairas et al. [37] reviewed methods and tools used for the building design optimisation while, more specifically, Kheiri [31] reviewed optimisation methods for building geometry and envelope design and Stevanović [90] reviewed work on optimisation of the passive solar design of buildings. Although different optimisation methods have been used in building energy optimisation, such as direct search, simulated annealing, particle swarm optimisation, harmony search and ant colony optimisation, they appear rather sporadically in the literature, while a large majority of building design optimisation studies rely on genetic algorithms. In this aspect, one might object that the building energy optimisation community is somewhat lagging behind other engineering communities, in which newer optimisation methods are more easily embraced and applied in research. One potential reason for this is the influence of Caldas and Norford's early example, and another that genetic algorithms tend to work well for building performance optimisation problems.

Genetic algorithms, inspired by biological evolution, work by randomly selecting an initial population of candidate solutions for the optimisation problem, and then evolve the population over a number of generations by repeated application of selection, reproduction, mutation and recombination, with the goal of improving candidates' values of the objective function(s). Gutowski [21] considered genetic algorithms from the linear algebraic viewpoint and reached some general recommendations on setting up their basic parameters as follows. Suppose that N_b bits are enough to describe values of all parameters of a particular solution, so the search space contains at most 2^{N_b} solutions. Recommended minimal population size is $\lceil \sqrt{2N_b} \rceil$, recommended mutation rate is $p_{\mathrm{mut}} \approx 1 - 0.82^{1/N_b}$ (so that only a few bits would be mutated in the whole population), while the crossover rate can be chosen between 0.7 and 1.0, independently of N_b. Such population is expected to become

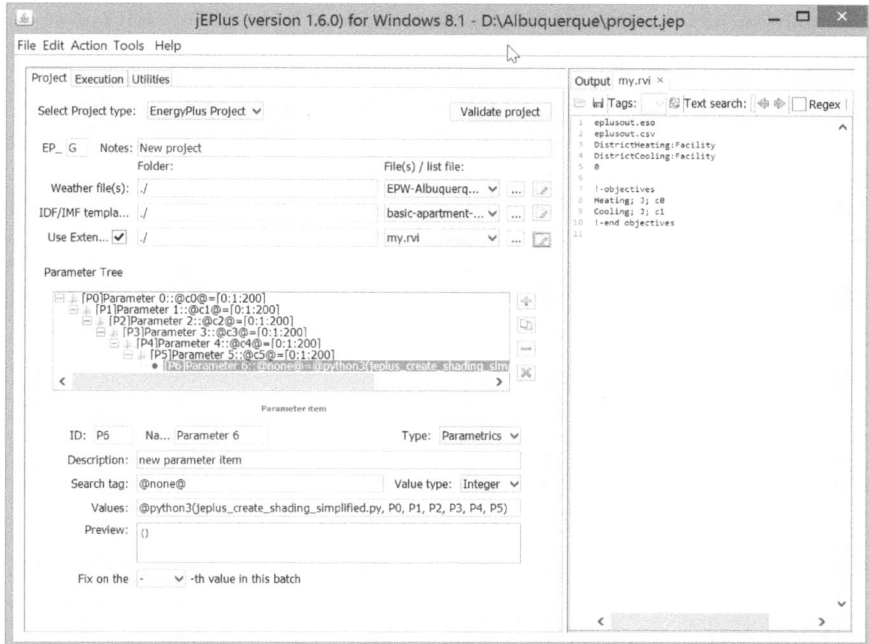

Fig. 3.2 A screenshot of the jEPlus main window

mature after evolving it for $1/2\,p_{mut}$ generations, and expected to produce its optimal solutions within 5–6 such epochs, so that the execution of the genetic algorithm can be stopped after $3/p_{mut}$ generations. The reader is referred to [21] for a detailed explanation of these recommendations.

Several software packages have appeared for the practical coupling of optimisation algorithms with building performance simulations, and we will briefly present here a combination of jEPlus [112–114] and jEPlus+EA [115]. jEPlus is an open-source software that enables performing parametric EnergyPlus simulations by describing a search space with sets of alternative values for specified simulation parameters and running simulations for either the whole search space or its representative sample. A screenshot of its main window is shown in Fig. 3.2.

A project in jEPlus first specifies, in the upper left part of the window, the weather file and the idf or imf template that will be used for subsequent EnergyPlus simulations. Values that will be collected from the simulation output are described in the rvi file, whose contents are shown in the right part of the screenshot. In this particular example, *DistrictHeating:Facility* and *DistrictCooling:Facility*, that are also declared as output meters in the imf template, are collected from the simulation results. These two meters are then named as *Heating* and *Cooling* in the *!-objectives* section of the rvi file, which will be later used by jEPlus+EA. Here, *c0* and *c1* are used internally by jEPlus to denote collected simulation output values.

Parameters that describe the search space are listed in the middle-left part of the jEPlus window. Here, parameters with ids *P0–P5* have values in the range from 0 to 200 with a step 1 (hence 201 different values), which in this particular example represents feasible depths of control points of a NURBS-lined overhang that was considered in [92] and that will be described later in Sect. 4.2. The parameter *P6* illustrates the ability of jEPlus to use Python for preprocessing simulation files. Its value *@python3(jeplus_create_shading_simplified.py, P0, P1, P2, P3, P4, P5)* specifies the Python file that will be called before the simulation takes place and the list of parameters whose values will be sent to the file. The beginning of the file *jeplus_create_shading_simplified.py* looks as follows:

```
import sys
import os

idd_filename=os.path.join(sys.argv[4], 'Energy+.idd')
idf_filename=os.path.join(sys.argv[2], 'in.idf')

param=sys.argv[3].split(',')
depths=[float(param[i]) for i in range(len(param))]
```

which translates parameters *P0–P5* from jEPlus into a list of depths available for further use in the Python file. jEPlus itself supplies the following arguments to the Python file:

- *sys.argv[1]* is the project base folder,
- *sys.argv[2]* is the output folder where files with collected simulation outputs are located,
- *sys.argv[3]* is the comma-delimited string containing values of parameters listed in the *@python3* call in jEPlus and
- *sys.argv[4]* is the folder containing the simulation program binaries.

The fourth and the second arguments provide paths to the EnergyPlus idd file and the idf input file within each simulation folder, which are needed to use *eppy* [57, 58], a scripting language for navigating, searching and modifying the EnergyPlus idf file. The actual parameter values are listed in *sys.argv[3]*, which needs to be parsed into a list of strings param and converted to a list of floats depths or other necessary data types first, after which the rest of the Python file builds that particular solution from the search space within the idf file and returns control to jEPlus which further calls EnergyPlus to simulate the built idf file.

jEPlus serves to define the search space and objective functions for an optimisation problem. While it may run EnergyPlus simulations for either a sample or all instances from the search space (see jEPlus website [114] for a description of its much wider capabilities), it does not implement any optimisation method per se. To actually run a genetic algorithm for the defined optimisation problem, one needs to use its sibling tool jEPlus+EA [115], which implements the non-dominated sorting genetic algorithm NSGA-II [14], one of the most well-known genetic algorithm variants.

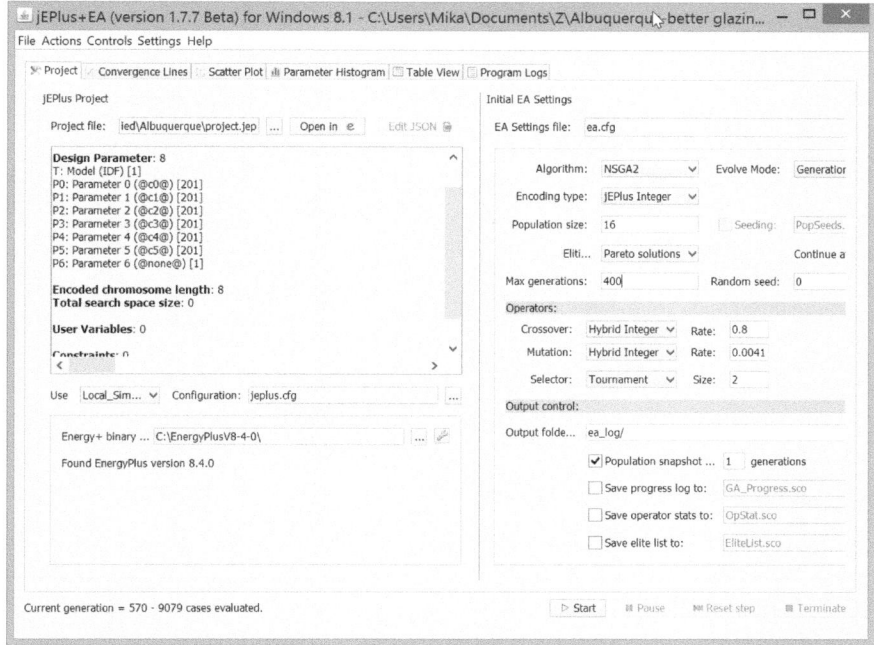

Fig. 3.3 A screenshot of the initial jEPlus+EA window

Figure 3.3 shows the initial tab of the jEPlus+EA window where the jEPlus project is validated in the left part, and genetic algorithm settings are specified in the right part.

Figure 3.4 shows the scatter plot tab of jEPlus+EA, depicting instances considered by the genetic algorithm so far as small circles with coordinates of the centre given by the respective *DistrictHeating:Facility* and *DistrictCooling:Facility* values. One should note here that Python preprocessing in jEPlus+EA is available from the version v1.7.7 beta only.

Another note to be taken care of is that when jEPlus+EA performs optimisation according to two or more objective functions, it has to update the Pareto front with each new generation. The size of the Pareto front will gradually increase with the number of generations, as NSGA-II is an elitist strategy that keeps the fittest candidates over different generations. Due to the quadratic worst-case time of the simple cull algorithm used to update the Pareto front, after several hundred generations jEPlus+EA may end up spending significantly more time updating the Pareto front and preparing the next generation than actually running EnergyPlus to simulate it, so one should aim for a compromise between Gutowski's recommendation to run genetic optimisation for at least $3/p_{mut}$ generations and own experience in practice.

Near-optimal solutions found by genetic optimisation may sometimes look rather unintuitive. One should note here that genetic optimisation is not guaranteed to

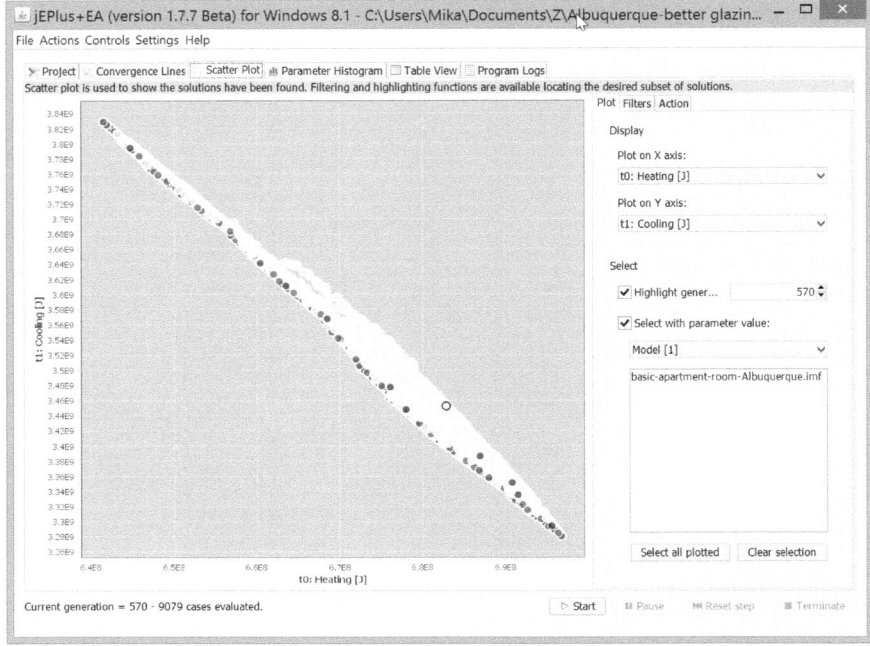

Fig. 3.4 A screenshot of the scatter plot tab of jEPlus+EA

find truly optimal solutions. This is hard to expect in cases when the search space contains billions of alternative solutions whose objective function values are placed in a relatively narrow range, implying that there might be thousands of alternatives that are within 0.1% of the true optimum. Rather, the true benefit of genetic optimisation is that it is able to consistently reach that 0.1% vicinity of the optimum values while simulating only a tiny part of the search space, while the researcher can then engage in further study of Pareto solutions in the search for their characterising properties.

After this brief introduction to genetic algorithms and their practical setup with jEPlus and jEPlus+EA, we now review the overhang shading studies in which genetic algorithms were prominently used.

Merla et al. [49] demonstrate usefulness and designer-friendliness of coupling of genetic optimisation with building performance simulations on a case study performed and analysed with a series of plugins for Rhinoceros:

- Grasshopper, graphical algorithm editor for Rhinoceros,
- Ladybug for importing and analysing weather data files,
- Honeybee for connecting Grasshopper with EnergyPlus, Radiance and DAYSIM, and
- Octopus for using SPEA2 genetic algorithm [116] from Grasshopper.

The main benefit of this setup is that the whole process takes place within a single software suite in which one defines model geometry, performs genetic optimisation and evaluates Pareto solutions. The case study aims to optimise depth of horizontal overhang, depth and inclination angle of vertical side fins and window-to-wall ratio for a three-storey office building in order to achieve a balance between the reduction of solar radiation incident on windows and the maximum internal daylight factor.

Amer and Wagdy [2] were interested in designing overhangs to enable proper daylighting for a case study of a south-facing cellular office in the desert climate of Cairo, Egypt, with abundant sunshine. Their goals were that the whole office has a sufficient amount of daylighting, with no part of the office overlit, in the sense that the illuminance levels at a given desk point are more than tenfold the target illuminance (300 lux) for more than 5% of the occupied hours. This cannot be done without shading in Cairo, as the office with a 5% window-to-wall ratio had only 9% of daylight autonomy, while with a 25% window-to-wall ratio it had 100% of daylight autonomy, but 43% of the office area was already overlit with 39% of annual sunlight exposure. As standard fixed shading devices allow some direct sunlight, which raise both overlit area and annual sunlight exposure above zero, their goal was to design an overhang shape that fully blocks direct solar radiation in a scholastically symmetric time period. Construction of such overhang, illustrated in Fig. 3.5, is achieved by using solar paths for a selected date: its outline consists of the part of the solar path from sunrise to noon seen from the eastern lower corner of the window and the part of the solar path from noon to sunset seen from the western lower corner of the window, connected by a straight line.

Genetic optimisation is performed by evolutionary solver Galapagos [78] for Rhinoceros, with the search space defined by several feasible values for the office window-to-wall ratio, outward protrusion of the shading device, shading reflectance and the date whose solar paths are used in the construction. The objectives were to maximise spatial daylight autonomy, the percentage of floor area in which 300 lux of illuminance is reached for at least 50% of occupied hours, and to minimise the percentage of overlit desk area and the annual sunlight exposure, the percentage of floor area in which 1000 lux of illuminance is reached for at least 250 h during occupied hours. Daylighting simulations were performed by DIVA-for-Rhino [25] for a grid of desk height points spaced 0.5 m apart and offset 0.25 m from the office walls. Selected Pareto optimal solutions are illustrated in Fig. 3.6. It can be seen that genetic optimisation selected solar paths of December 21 for overhang constructions, thus effectively blocking all direct solar radiation from entering the office. Large window-to-wall ratios were then necessary to let in enough ground reflected and diffuse solar radiation, which ensured 100% of spatial daylight availability and 0% of annual sunlight exposure.

The study of Sghiouri et al. [80], mentioned already in Sect. 3.2.1, used the combination of jEPlus+EA and TRNSYS for a single-objective optimisation of overhang lengths in four different thermal zones of a two-storey two-family building with two façades, located in three representative Moroccan cities of Casablanca, Marrakech and Oujda, with the goal of minimising discomfort degree hours. An interesting aspect of the study was that overhang lengths were optimised in two stages, in order

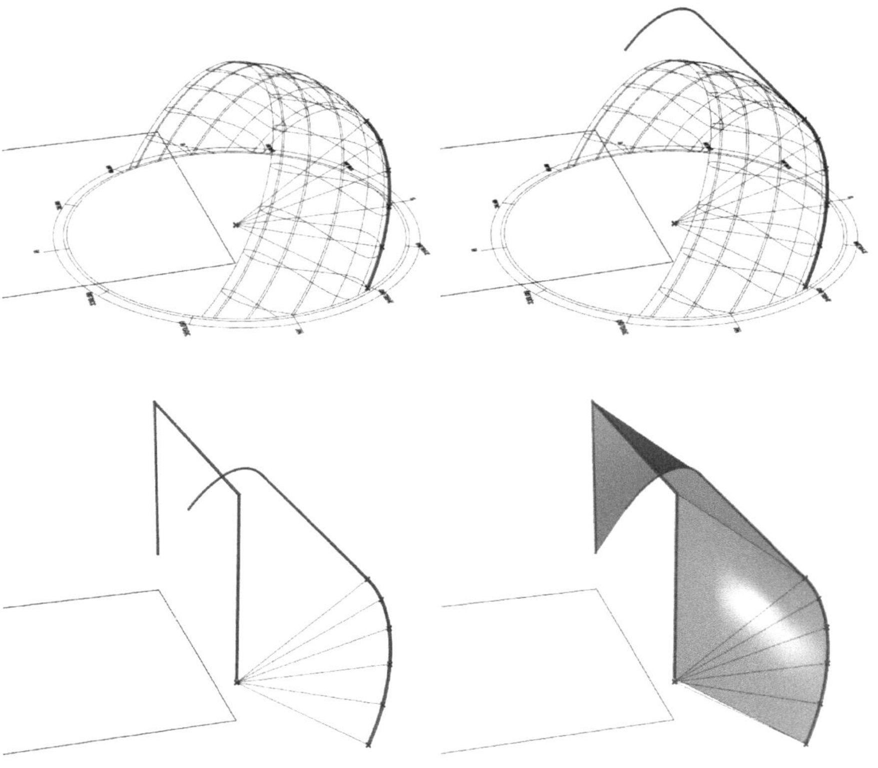

Fig. 3.5   Illustration of shading construction from [2]

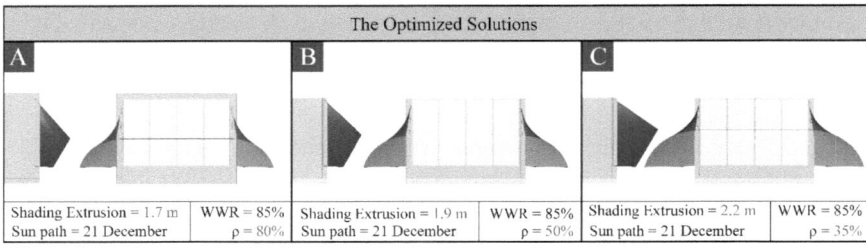

The Optimized Solutions					
A		B		C	
Shading Extrusion = 1.7 m	WWR = 85%	Shading Extrusion = 1.9 m	WWR = 85%	Shading Extrusion = 2.2 m	WWR = 85%
Sun path = 21 December	ρ = 80%	Sun path = 21 December	ρ = 50%	Sun path = 21 December	ρ = 35%

Fig. 3.6   Illustration of optimal shading devices from [2]

to narrow down the search space. The first stage was a rough optimisation that varied the lengths by 3 cm, that was used to suggest maximum feasible lengths of overhangs, while the second stage was a finer optimisation that varied their lengths by 1 cm to obtain optimal lengths more accurately.

Manzan and his coauthors [41–44] used genetic algorithms to optimise geometric parameters of an overhang in the form of a flat panel, with maximum depth of 2 m,

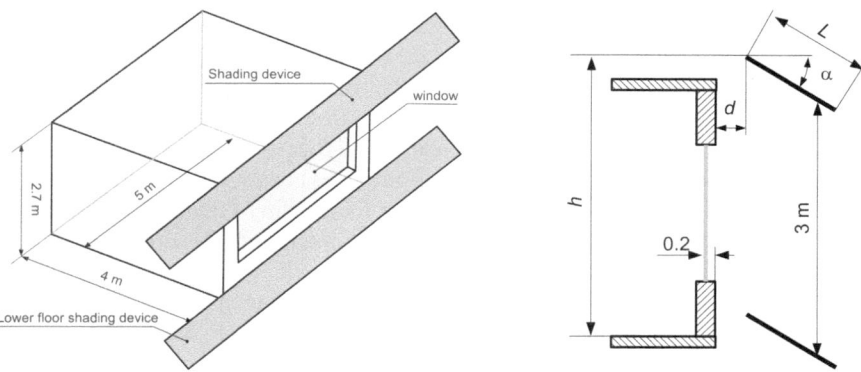

Fig. 3.7 Cellular office with an overhang and its geometric parameters [41]

positioned parallel to the window of a south-facing cellular office in Trieste, Italy, and inclined by its horizontal axis (see Fig. 3.7). The optimal overhang is defined as the one that minimises annual primary energy consumption, given as a linear combination of loads for heating, cooling and artificial lighting. Effects of two different glazing systems, the standard double glass and a high performance glass with a small g value, and the presence of 0.2 m window reveals have been considered through separate optimisations performed for different cases based mostly on the glass type and the presence of window reveals. Office energy performance was simulated with ESP-r [16], while genetic optimisation was managed by modeFRONTIER [18], with each optimisation performed on a population of 16 individuals over 100 generations. The search space was defined by varying four overhang parameters: height h, depth L, inclination angle α and distance d from the wall (see Fig. 3.7). Parameters were not fully independent, as overhang was assumed to have limited horizontal protrusion and not to obstruct the view from the office.

Main differences between optimisations performed in [41–44] lie in the treatment of lighting within the cellular office. Lighting loads in [44] have been computed with Radiance [62], with artificial lighting switched on when illuminance is below 300 lux and switched off when it is above 450 lux. A substantial reduction of primary energy consumption was obtained with a deeper shading panel, with a rather horizontal deep overhang high above the window selected as optimal in most cases. Due to the already strong shading characteristics of the high performance glass, the optimal depth of the overhang in cases with high performance glass was always smaller than in cases with standard double glass. In all cases, optimisation results suggest that artificial lighting load has to be taken into account in the design of energy-efficient shading devices.

Lighting loads in [41] have been computed by DAYSIM [70], a Radiance-based daylight simulator, with artificial lighting dimmed until the illuminance at the centre of the room at desk height reaches the minimum threshold of 500 lux. Besides Trieste, optimisation has been performed for the location of Rome, Italy, as well, where

Rome has almost twice as much annual global horizontal radiation as Trieste, and separately for narrow (maximum depth of 1 m) and broad overhangs (maximum depth of 2 m). For broader overhangs (that have also been considered in [44]), optimal solutions are nearly horizontal in Trieste and in Rome for standard double glass, while high performance glass in Rome leads to higher optimal inclination angles. For narrower overhangs, the optimal inclination angles are always over 30° in Trieste, and even higher in Rome, in order to increase the shading effect of the overhang. Manzan has also computed the useful daylight illuminance UDI along a line at the centre of the office from the window to the opposite wall at desk height. While high performance glass offered better daylight distribution than standard double glass, the area in proximity of the window received too much daylight with low values of UDI in each case, indicating that for a comfortable workspace an additional moveable shading device should be installed to avoid glare problems.

In Manzan and Padovan [43], only high performance glass has been considered with glare protection provided by the automated internal Venetian blind, set up to be fully lowered as soon as direct solar irradiance reaches $50\,W/m^2$ on either of two interior sensors (and fully open otherwise). This time, two different office orientations, south and south-west, have been considered, while the panel distance d from the wall has been set to zero, so only the values of h, L and α have been optimised. In addition to the annual primary energy consumption, another objective function has been added: the number of hours of activation of the automated internal Venetian blind. Minimisation of this objective should ensure optimal levels of daylight during the year and an unobstructed view outside the window. For both south and south-west facing offices, three particular overhang configurations from the Pareto front, the one with the smallest primary energy consumption, the one with the smallest number of hours of blind activation and an intermediate Pareto solution, have been analysed in detail. All six Pareto solutions have a deeper overhang (1.71–2.00 m) and higher inclination angles (20.7°–34.3°), reducing primary energy consumption for 18.1–23.9% compared to the case without an overhang panel. Automatic Venetian blinds are activated mostly during winter months for the south-facing office. In the case of the south-west facing office with a shallow overhang panel, they are activated throughout the year. As expected, the overhang panel has reduced thermal performance for the south-west facing office, with higher cooling loads due to its inability to shade lower solar positions (one of the constraints is that the overhang does not obstruct the view outside). Nevertheless, optimisation results show that highly obstructive overhangs do not achieve the lowest energy consumption.

Manzan and Clarich [42] consider the south-facing office only with high performance glass. Both double low-e and triple low-e are simulated, but they turn out to yield very similar results. In this case, the internal Venetian blinds are assumed to be manually controlled and set as either fully open, or fully lowered with horizontal slats, or fully lowered with slats inclined at 45°, which is simulated with DAYSIM through a user behaviour control model Lightswitch [69]. The panel distance d from the wall has been set to zero again, so that just the values of h, L and α have been optimised, with two objective functions: the annual primary energy for heating, cooling and artificial lighting and the annual number of hours with internal blinds deployed at 45° angle during the occupancy time, thus looking for solutions that minimise the time with obstructed outside view. Overall conclusions are relatively similar to previous studies: optimal overhang panels reduce the primary energy consumption for approximately 24% for double glazing and 26.5% for triple glazing; smaller panels lead to excessive daylight near the window but to better useful daylight illuminance deeper in the office; blinds are operated mostly during winter when the Sun has lower altitude, showing that the overhang reduces the time when direct solar radiation forces occupants to incline the blinds.

What is, however, more interesting about the last two studies [42, 43] is the way in which Pareto front solutions were obtained. Namely, DAYSIM simulations, based on raytracing techniques, take up a significant amount of time (several minutes on a quad-core processor per simulated case), which prevents simulation of an excessive number of cases often needed to approximate the Pareto front with genetic optimisation. Instead, current simulation results are used to obtain a statistical model to predict the results of future simulations, to find optimal solutions of the statistical model by genetic optimisation, and to actually simulate just those expected optimal solutions, after which the process is repeated by updating the statistical model with new simulation results, enabling identification of the Pareto front with far less computational resources. Statistical models that approximate and mimic the behaviour of the original simulation model are often called *surrogate models*. According to [53], they are usually constructed as follows:

- Create a sample of inputs and simulate corresponding model variants, whose outputs form a database for training a surrogate model,
- Construct the surrogate model from the output database by an appropriate method (artificial neural network, response surface, support vector machine, etc.) and
- Validate the obtained surrogate model;

where the last two steps are usually reiterated until a satisfactory surrogate model is attained. Both [42, 43] use adaptive response surface metamodels [11, 59], as implemented in the FAST algorithm of modeFRONTIER [18]. Manzan and Padovan in [43] used just 150 real simulations in 10 iteration steps of 15 overhang panel designs each to approximate the Pareto front. Manzan and Clarich [42] further compared Pareto fronts obtained by the response surface metamodels in FAST and the classical NSGA-II genetic algorithm, observing that they are very similar in both cases and that after 15-20 iterations there are no great differences between them.

Ekici et al. [15] used artificial neural networks to develop surrogate models for spatial daylight autonomy $sDA_{300lx,50\%}$ and annual sunlight exposure $ASE_{1000lx,250h}$ in a high-rise building in a dense urban district. To obtain more representative daylighting results, the high-rise building is divided into five vertical zones. The free variables are the overhang length and the glazing type for each façade in each zone. Using them, 500 uniformly distributed building samples are generated and simulated in each zone using DIVA-for-Rhino [25]. These simulation results were then used as the training data, where a separate artificial neural network with a single hidden layer of 20 neurons was trained using backpropagation with bipolar sigmoid activation function for each of the five zones and each of the two daylight performance indicators. These ten neural networks achieved a correlation coefficient of $R^2 = 98.25\%$, verified against additional DIVA simulations for building models with similar design parameters. Finally, the single-objective self-adaptive differential evolution algorithm [8] was used to maximise $sDA_{300lx,50\%}$ subject to the constraint that $ASE_{1000lx,250h} \leq 20\%$, using the trained neural networks instead of DIVA simulations to compute the sDA and ASE values.

Harmony search [20, 40, 97] is another optimisation approach used in a recent overhang shading study. The approach is envisaged to mimic the process of improvisation that jazz musicians use to achieve the best sounding harmony. Each "musician" here represents one of the parameters in the optimisation problem, while the feasible values represent the pitches of their instruments. The process maintains a harmony memory of solutions used so far, which are initially generated uniformly at random. The process uses two probabilistic parameters: the harmony search consideration rate $HCMR$ and the pitch adjustment rate PAR that control selection of parameter values for a new solution. To generate a new solution, each parameter with probability $HCMR$ selects its value, uniformly at random, from corresponding values in the harmony memory, after which with probability PAR it selects a feasible value adjacent to the value selected from the harmony memory. Otherwise (with probability $1 - HCMR$), the parameter takes a random feasible value. After the new solution is generated in this way, it is simulated to obtain its objective function values and, if it is better, it replaces the worst solution in the harmony memory, with the best solution returned after the maximum number of iterations. Although relatively widely used as a separate optimisation approach in engineering disciplines, earlier studies [104, 105] have shown that harmony search is actually a special case of evolution strategies, one of the most classic search heuristics [68].

Khoroshiltseva et al. [32] use harmony search to investigate the appropriate design of overhang and side fins for existing residential social housing in Madrid, Spain, built in the 1950s that would simultaneously optimise several conflicting objectives: reduce the overheating period within the building, minimise primary energy needs for heating and lighting and have a small surface area. Shading device for each window is composed of six elements, with overhang and each fin divided into two rectangular parts as shown in Fig. 3.8, whose depths are then optimised. The case study simultaneously optimises such shading devices surrounding two south-facing windows and two west-facing windows, assuming that windows with the same orientation are surrounded by the same shading device. Hence, the south-facing and the west-facing shading devices are described by a total of 12 depth values, with each value in the set {0 m, 0.05 m, ..., 0.65 m, 0.7 m}, leading to the extremely large search space of 15^{12} feasible solutions. However, solutions where the fin elements are deeper than the overhang elements or where two adjacent overhang elements differ by more than 0.1 m are deemed infeasible for aesthetic reasons.

Optimisation was done by combining harmony search, with $HCMR = 0.9$ and $PAR = 0.8$, and information achieved by statistical models [5, 7, 86, 87], in order to obtain good approximation of the Pareto front in 1500 simulations only. Overheating here is defined as the proportion of hours during the year when the room temperature exceeds 26 °C. Discussion of a number of Pareto solutions that have feasible shape

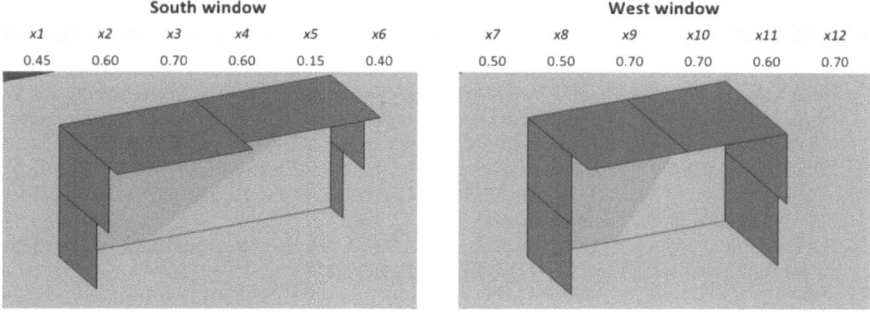

Fig. 3.8 Structure of overhang and side fins considered in [32]

and large effect on overheating shows that the size of their fins is generally smaller for southern windows than for western windows, while the depths of overhang elements and southern fins at western windows are almost always set to the maximum possible value (0.7 m). While overhang and side fin elements cannot eliminate overheating periods, they still manage to reduce their length from 21.3% down to 16.9–18.0%.

Two further overhang studies [91, 92] in which genetic optimisation was prominently used will be reviewed in the next chapter, as they deal with overhangs that are not bounded to rectangular shapes.

References

1. Acosta I, Muñoz C, Esquivias P, Moreno D, Navarro J (2015) Analysis of the accuracy of the sky component calculation in daylighting simulation programs. Sol Energy 119:54–67
2. Amer M, Wagdy A (2016) Multivariable optimization for zero over-lit shading devices in hot climate. In: Proceedings of the 3rd IBPSA-England conference BSO 2016, Newcastle, UK. Accessed 12–14 Sep 2016
3. ASHRAE (2005) Handbook: fundamentals. ASHRAE, Atlanta
4. Baker N (2000) We are all outdoor animals. In: Steemers K, Yannas S (eds) Proceedings of PLEA 2000: architecture, city, environment, Cambridge, United Kingdom, 2000 Jul 2–5. James & James (Science Publishers), London, pp 553–555
5. Baragona R, Battaglia F, Poli I (2011) Evolutionary statistical procedures: an evolutionary computation approach to statistical procedures design and applications. Springer, Berlin
6. Bigladder Software. Shading module: engineering reference—EnergyPlus 8.9. https://bigladdersoftware.com/epx/docs/8-9/engineering-reference/shading-module.html. Accessed Feb 2022
7. Borrotti M, March DD, Slanzi D, Poli I (2014) Designing lead optimization of MMP-12 inhibitors. Comput Math Methods Med 2014, Article ID 258627
8. Brest J, Zumer V, Maucec MS (2006) Self-adaptive differential evolution algorithm in constrained real-parameter optimization. In: Proceedings of the 2006 IEEE international conference on evolutionary computation, Vancouver, Canada, 2006 Jul 16–21. IEEE, pp 215–222
9. Caldas LG, Norford LK (2003) Genetic algorithms for optimisation of building envelopes and the design and control of HVAC systems. J Sol Energ-T ASME 125:343–351

10. Central building research institute (1987) Handbook of functional requirements of buildings (other than industrial buildings). SP: 41 (S& T). Bureau of Indian Standard, New Delhi
11. Clarich A, Pediroda V, Poloni C (2005) A fast and robust adaptive methodology for design under uncertainties based on DACE response surface and game theory. In: Annicchiarico W, Priaux J, Cerrolaza M, Winter G (eds) Proceedings of evolutionary algorithms and intelligent tools in engineering optimization. Cimne, Barcelona, pp 172–181
12. Daneshyar M (1979) Solar radiation statistics for Iran. Sol Energy 21:345–349
13. David M, Donn M, Garde F, Lenoir A (2011) Assessment of the thermal and visual efficiency of solar shades. Build Environ 46:1489–1496
14. Deb K, Pratap A, Agarwal S, Meyarivan T (2002) A fast and elitist multiobjective genetic algorithm: NSGA-II. IEEE T Evol Comput 6(2):182–197
15. Ekici B, Kazanasmaz T, Turrin M, Tasgetiran MF, Sariyildiz IS (2019) A methodology for daylight optimisation of high-rise buildings in the dense urban district using overhang length and glazing types with surrogate modelling. In: Scartezzini JL, Smith B (eds) Proceedings of CISBAT 2019: climate resilient cities—energy efficiency & renewables in the digital era, Lausanne, Switzerland, 2019 Sep 4–6. J Phys Conf Ser 2019; 1343:012133
16. Energy systems research unit of the University of Strathclyde. ESP-r. https://esru.strath.ac.uk/applications/esp-r/. Accessed Feb 2022
17. Esquivias PM, Munoz CM, Acosta I, Moreno D, Navarro J (2016) Climate-based daylight analysis of fixed shading devices in an open-plan office. Lighting Res Technol 48(2):205–220
18. Esteco SpA (2022) ModeFrontier: process automation and optimization in the engineering design process. https://esteco.com/modefrontier. Accessed Feb 2022
19. Fortune S (1989) Stable maintenance of point set triangulations in two dimensions. In: Proceedings of the 30th annual IEEE symposium on foundations of computer science, Research Triangle Park, USA, 1989 Oct 30–Nov 1. IEEE, Washington, DC, pp 494–499
20. Geem ZW, Kim JH, Loganathan GV (2001) A new heuristic optimization algorithm: harmony search. Simulation 76:60–68
21. Gutowski MW (2005) Biology, physics, small worlds and genetic algorithms. In: Shannon S (ed) Leading edge computer science research. Nova Science Publishers, Hauppauge, pp 165–218
22. Hopkinson RG (1972) Glare from daylighting in buildings. Appl Ergon 3:206–215
23. Hviid CA, Nielsen TR, Svendsen S (2008) Simple tool to evaluate the impact of daylighting on building energy consumption. Sol Energy 82(9):787–798
24. Illuminating Engineering Society (2012) IES approved method: spatial daylight autonomy (sDA) and annual sunlight exposure (ASE). (IES LM-83-12). Illuminating Engineering Society, New York
25. Jakubiec A, Reinhart C (2011) Diva 2.0: integrating daylight and thermal simulation using Rhinoceros 3D, Daysim and EnergyPlus. In: Proceedings of building simulation 2011: the 12th international IBPSA conference, Sydney, Australia, 2011 Nov 14–16. IBPSA, pp 2202–2209
26. Jones RE Jr (1980) Effects of overhang shading of windows having arbitrary azimuth. Sol Energy 24:305–312
27. Jones NL, Greenberg DP, Pratt KB (2012) Fast computer graphics techniques for calculating direct solar radiation on complex building surfaces. J Build Perform Simu 5:300–312
28. Liu BYH, Jordan C (1961) Daily insolation on surfaces tilted toward the equator. ASHRAE J 3:53–59
29. Kabre C (1999) Winshade: a computer design tool for solar control. Build Environ 34:263–274
30. Källblad K (1999) A method to estimate the shading of solar radiation theory and implementation in a computer program. In: Proceedings of building simulation 1999: the 6th international IBPSA conference, Kyoto, Japan, 1999 Sep 13–15. IBPSA, 1999, cód. A-23
31. Kheiri F (2018) A review on optimisation methods applied in energy-efficient building geometry and envelope design. Renew Sust Energ Rev 92:897–920
32. Khoroshiltseva M, Slanzi D, Poli I (2016) A Pareto-based multi-objective optimization algorithm to design energy-efficient shading devices. Appl Energ 184:1400–1410

33. Kirimtat A, Koyunbaba BK, Chatzikonstantiou I, Sariyildiz S (2016) Review of simulation modeling for shading devices in buildings. Renew Sust Energ Rev 53:23–49
34. Klein SA (1977) Calculation of monthly average insolation on tilted surfaces. Sol Energy 19:325–329
35. Lim HS, Kim G (2010) Predicted performance of shading devices for healthy visual environment. Indoor Built Environ 19:486–496
36. Loonen RCGM, Hensen JLM (2013) Dynamic sensitivity analysis for performance-based building design and operation. In: Wurtz E (ed) Proceedings of building simulation 2013: the 13th international conference of IBPSA, Chambery, France, 2013 Aug 26–28. IBPSA, Toronto, pp 299–305
37. Machairas V, Tsangrassoulis A, Axarli K (2014) Algorithms for optimisation of building design: a review. Renew Sust Energ Rev 31:101–112
38. Maestre IR, Blázquez JLF, Gallero FJG, Cubillas PR (2015) Influence of selected solar positions for shading device calculations in building energy performance simulations. Energ Build 101:144–152
39. Maestre IR, Pérez-Lombard L, Foncubierta JL, Cubillas PR (2013) Improving direct solar shading calculations within building energy simulation tools. J Build Perform Simu 6:437–448
40. Mahdavi M, Fesanghary M, Damangir E (2007) An improved harmony search algorithm for solving optimization problems. Appl Math Comput 188:1567–1579
41. Manzan M (2014) Genetic optimization of external fixed shading devices. Energ Build 72:431–440
42. Manzan M, Clarich A (2017) FAST energy and daylight optimization of an office with fixed and movable shading devices. Build Environ 113:175–184
43. Manzan M, Padovan R (2015) Multi-criteria energy and daylighting optimization for an office with fixed and movable shading devices. Adv Build Energ Res 9:238–252
44. Manzan M, Pinto F (2009) Genetic optimization of external shading devices. In: Proceedings of building simulation 2009: the 11th international IBPSA conference, Glasgow, Scotland, 2009 Jul 27–30. IBPSA, pp 180–187
45. Mardaljević J (2003) Precission modelling of parametrically defined solar shading systems: pseudo-Changi. In: Proceedings of building simulation 2003: the 8th international IBPSA conference, Eindhoven, Netherlands, 2003 Aug 11–14. IBPSA, pp 823–830
46. Marsh A (2005) The application of shading masks in building simulation. In: Proceedings of building simulation 2005: the 9th international IBPSA conference, Montréal, Canada, 2005 Aug 15–18. IBPSA, pp 725–732
47. McNeil A, Burrell G (2016) Applicability of DGP and DGI for evaluating glare in a brightly daylit space. In: Proceedings of SimBuild 2016: ASHRAE and IBPSA-USA building performance modeling conference, Salt Lake City, USA, 2016 Aug 8–12. ASHRAE, Atlanta, pp 57–64
48. Mendes N, Oliveira RCLF, Henrique G (2003) Domus 2.0: a whole building hygrothermal simulation program. In: Proceedings of building simulation 2003: the 8th international IBPSA conference, Eindhoven, Netherlands, 2003 Aug 11–14. IBPSA, pp 863–870
49. Merla M, Diaferia R, Dibari G (2016) Application of parametric study and generative algorithms to optimize building physics analyses. In: Roche PL, Schiler M (eds) Proceedings of PLEA 2016: the 32nd international conference on passive and low energy architecture, Los Angeles, 2016 Jul 11–13, pp 23–28
50. Messac A (1996) Physical programming: effective optimization for computational design. AIAA J 34(1):149–158
51. Nabil A, Mardaljević J (2005) Useful daylight illuminance: a new paradigm for assessing daylight in buildings. Lighting Res Technol 37:41–57
52. Nabil A, Mardaljević J (2006) Useful daylight illuminances: a replacement for daylight factors. Energ Build 38:905–913
53. Nguyen AT, Reiter S, Rigo P (2014) A review on simulation-based optimization methods applied to building performance analysis. Appl Energ 113:1043–1058

54. Nielsen TR (2005) Simple tool to evaluate energy demand and indoor environment in the early stages of building design. Sol Energy 78(1):73–83
55. Niewienda A, Heidt FD (1996) Sombrero: a PC-tool to calculate shadows on arbitrarily oriented surfaces. Sol Energy 58:253–263
56. Perez R, Ineichen P, Seals R, Michalsky J, Stewart R (1990) Modelling daylight availability and irradiance components from direct and global irradiance. Sol Energy 44:271–289
57. Philip S (2022) Scripting language for EnergyPlus. https://github.com/santoshphilip/eppy/. Accessed Feb 2022
58. Philip S (2022) Welcome to eppy's documentation. https://eppy.readthedocs.io/en/latest/. Accessed Feb 2022
59. Poloni C, Pediroda V, Clarich A (2005) A fast and robust adaptive methodology for design under uncertainties based on DACE response surface and game theory. ERCOFTAC Ser 66:29–36
60. Preparata FP, Shamos MI (1985) Computational geometry-an introduction. Springer, New York
61. Queiroz N, Westphal FS, Pereira FOR (2020) A performance-based design validation study on EnergyPlus for daylighting analysis. Sol Energy 183:107088
62. Radiance—Radsite. https://radiance-online.org. Accessed Feb 2022
63. Raeissi S, Taheri M (1998) Optimum overhang dimensions for energy saving. Build Environ 33:293–302
64. Rahman AN, Mizanur ANM, Satyamurty VV (1999) Overhang shading factor values for windows of general azimuth angle evaluated under extraterrestrial and terrestrial conditions. Int J Energ Res 23:235–245
65. Ramos G, Ghisi E (2010) Analysis of daylighting calculated using the EnergyPlus programme. Renew Susta Energ Rev 14(7):1948–1958
66. Rao S, Tzempelikos A (2010) The impact of exterior overhangs on the daylighting performance of office spaces. In: Proceedings of the 1st international high performance building conference, West Lafayette, USA. Accessed 12–15 July 2010
67. Rea MS (ed) (2000) The IESNA lighting handbook: reference & application. Illuminating Engineering Society of North America, New York
68. Rechenberg I (1973) Evolutionsstrategie: Optimierung technischer systeme nach prinzipien der biologischen evolution. Stuttgart: frommann-holzboog Verlag
69. Reinhart CF (2004) Lightswitch-2002 a model fo manual and automated control of electric lighting and blinds. Sol Energy 77:15–28
70. Reinhart C (2022) DAYSIM: advanced daylight simulation software. https://github.com/MITSustainableDesignLab/Daysim. Accessed Feb 2022
71. Reinhart C, Mardaljević J, Rogers Z (2006) Dynamic daylight performance metrics for sustainable building design. Leukos 3:1–25
72. Robbins CL (1986) Daylighting-design and analysis. Van Nostrand Reinhold Company, New York
73. Rocha APdA, Goffart J, Houben L, Mendes N (2016) On the uncertainty assessment of incident direct solar radiation on building facades due to shading devices. Energ Buildings 133:295–304
74. Rocha APdA, Oliveira RCLF, Mendes N (2017) Experimental validation and comparison of direct solar shading calculations within building energy simulation tools: polygon clipping and pixel counting techniques. Sol Energy 158:462–473
75. Rocha APdA, Reynoso-Meza G, Oliveira RCLF, Mendes N (2020) A pixel counting based method for designing shading devices in buildings considering energy efficiency, daylight use and fading protection. Appl Energ 262:114497
76. Roche L (2002) Summertime performance of an automated lighting and blinds control system. Lighting Res Technol 34:11–25
77. Roche L, Dewey E, Littlefair P (2000) Occupant reactions to daylight in offices. Lighting Res Technol 32:119–126

78. Rutten D (2022) Evolutionary principles applied to problem solving. https://grasshopper3d. com/profiles/blogs/evolutionary-principles. Accessed Feb 2022
79. Sameti M, Jokar MA (2017) Numerical modelling and optimization of the finite-length overhang for passive solar space heating. Intell Build Int 9:204–221
80. Sghiouri H, Mezrhab A, Karkri M, Naji H (2018) Shading devices optimization to enhance thermal comfort and energy performance of a residential building in Morocco. J Build Eng 18:292–302
81. Shafavi NS, Tahsildoost M, Zomorodian ZS (2020) Investigation of illuminance-based metrics in predicting occupants' visual comfort (case study: architecture design studios). Sol Energy 197:111–125
82. Shafavi NS, Zomorodian ZS, Tahsildoost M, Javadi M (2020) Occupants visual comfort assessments: a review of field studies and lab experiments. Sol Energy 208:249–274
83. Sharp K (1982) Calculation of monthly average insolation on a shaded surface at any tilt and azimuth. Sol Energy 28:531–538
84. Shaviv E, Yezioro A (1997) Analyzing mutual shading among buildings. Sol Energy 59:83–88
85. Siegel R, Howell JR (1972) Thermal radiation heat transfer. McGraw Hill, New York
86. Slanzi D, Lucrecia DD, Poli I (2015) Querying bayesian networks to design experiments with application to 1AGY serine esterase protein engineering. Chemometr Intell Lab Syst 149(A):28–38
87. Slanzi D, Poli I (2014) Evolutionary bayesian network design for high dimensional experiments. Chemometr Intell Lab Syst 135:172–182
88. Stevanović S (2018) Urban planning indicators in sustainable building assessment methods. J Eng Res 6:75–88
89. Stevanović S (2016) Parametric study of a cost-optimal, energy efficient office building in Serbia. Energy 117:492–505
90. Stevanović S (2013) Optimization of passive solar design strategies: a review. Renew Susta Energ Rev 25:177–196
91. Stevanović S, Stevanović D (2018) Optimisation of curvilinear external shading of windows in cellular offices. PLoS ONE 13:e0203575
92. Stevanović S, Stevanović D, Dehmer M (2019) On optimal and near-optimal shapes of external shading in apartment buildings. PLoS ONE 14:e0212710
93. Sutherland IE, Hodgman GW (1974) Reentrant polygon clipping. Commun ACM 17:32–42
94. Tappeta RV, Renaud JE, Messac A, Sundararaj GJ (2000) Interactive Physical Programming: Tradeoff Analysis and Decision Making in Multicriteria Optimization. AIAA J 38(5):917–926
95. Turan I, Chegut A, Fink D, Reinhart C (2020) The value of daylight in office spaces. Build Environ 168:106503
96. Utzinger DM, Klein SA (1979) A method of estimating monthly average solar radiation of shaded receivers. Sol Energy 23:369–378
97. Valian E, Tavakoli S, Mohanna S (2014) An intelligent global harmony search approach to continuous optimization problems. Appl Math Comput 232:670–684
98. Vanthoor BHE, Henten EJV, Stanghellini C, Visser PHBD (2011) A methodology for model-based greenhouse design: part 3, sensitivity analysis of a combined greenhouse climate-crop yield model. Biosyst Eng 110:396–412
99. Vasconcellos GDd (2017) Evaluation of annual sunlight exposure (ASE) as a proxy to glare: a field study in a NZEB and LEED certified office in San Francisco. MS thesis. University of California, Berkeley
100. Vatti BR (1992) A generic solution to polygon clipping. Commun ACM 35:56–63
101. Waldram PJ (1925) The natural and artificial lighting of buildings. J Roy Inst Brit Archit 32(405–426):441–446
102. Warg LG, Shakespeare R (1998) Rendering with Radiance: the art and science of lighting visualization. Morgan Kaufmann, San Francisco
103. Weiler K, Atherton P (1977) Hidden surface removal using polygon area sorting. In: SIGGRAPH '77 Proceedings of the 4th annual conference on Computer graphics and interactive techniques, San Jose, California, 1977 Jul 20–22. ACM, New York, pp 214–222

104. Weyland D (2010) A rigorous analysis of the harmony search algorithm: how the research community can be misled by a "novel" methodology. Int J Appl Metaheuristic Comput 1:50–60
105. Weyland D (2015) A critical analysis of the harmony search algorithm-How not to solve sudoku. Operat Res Perspect 2:97–105
106. Wienold J, Christoffersen J (2006) Evaluation methods and development of a new glare prediction model for daylight environments with the use of CCD cameras. Energ Buildings 38:743–757
107. Włodarczyk D, Nowak H (2009) A simple method of determining the influence of the overhang on window solar gains. In: Proceedings of building simulation 2009: the 11th international IBPSA conference, Glasgow, Scotland, 2009 Jul 27–30. IBPSA, pp 1617–1622
108. Yanda RF, Jones RE Jr (1983) Shading effects of finite width overhang on windows facing toward the equator. Sol Energy 30:171–180
109. Yener AK (1999) A method of obtaining visual comfort using fixed shading devices in rooms. Build Environ 34:285–291
110. Yezioro A, Shaviv E (1994) Shading: a design tool for analyzing mutual shading between buildings. Sol Energy 52:27–37
111. Yu F, Wennersten R, Leng J (2020) A state-of-art review on concepts, criteria, methods and factors for reaching "thermal-daylighting balance." Build Environ 186:107330
112. Zhang Y, Korolija I (2010) Performing complex parametric simulations with jEPlus. In: Proceedings of SET2010, 9th international conference on sustainable energy technologies, Shanghai, China, 2010 Aug 24–27. Shanghai Jiao Tong University
113. Zhang Y (2009) 'Parallel' EnergyPlus and the development of a parametric analysis tool. In: Proceedings of building simulation 2009: 11th international IBPSA conference, Glasgow, United Kingdom, Jul 27–30. IBPSA, pp 1382–1388
114. Zhang Y, Korolija I (2022) jEPlus—an EnergyPlus simulation manager for parametrics. https://jeplus.org. Accessed Feb 2022
115. Zhang Y, Korolija I (2022) jEPlus+EA User's Guide. https://jeplus.org/wiki/doku.php?id=docs:jeplus_ea:start. Accessed Feb 2022
116. Zitzler E, Laumanns M, Thiele L (2001) SPEA2: improving the strength Pareto evolutionary algorithm. TIK-Report 103. Zürich: ETH. https://doi.org/10.3929/ethz-a-004284029. Accessed Feb 2022

Chapter 4
Design Methods for Particular Overhang Types

Abstract This chapter deals with overhang studies that do not easily fit the previous two chapters due to their innovative approaches. The first two sections bring the methods motivated by Kaftan's remark that one should not expect optimal overhang shape to be of simple geometry even for a rectangular window, as solar gains and daylight intensities vary with changing solar angles. Methods in the first section are based on the division of the underlying overhang support surface into a cell array which are then treated individually, while the second section discusses a method that divides the support surface into a one-dimensional array of strips attached to the wall, whose outer edges are modelled by a smooth, continuous NURBS line. The third section presents a method in which the shading design problem is reversed so that instead the window is trimmed to optimally suit a given rectangular overhang. The fourth section discusses the results of a few studies that recommend overhangs as suitable exterior shading devices for building retrofits due to their structural properties. The last section describes methods for constructing and optimising PV integrated overhangs, whose sunny side uses blocked direct solar radiation to generate electricity, and movable overhangs, whose position can be updated from season to season.

Keywords Shading support surface · NURBS · Window fitting · Shading retrofit · Movable overhangs · Photovoltaic overhangs.

4.1 Subdividing Shading Support Surface

Motivated by the expectation that the optimal overhang shape for a rectangular window need not be rectangular itself, Kaftan [9] developed the *cellular method* that divides the overhang support surface into a two-dimensional array of cells and examines each cell individually for the necessity of its inclusion in the final overhang structure. The shading necessity of each cell is calculated separately for each hour of a given period, with the hourly necessities summed together to yield the overall shading necessity for the whole period. The hourly shading necessity of a cell is determined by taking into account whether direct solar radiation passing through the cell during the given hour reaches the window and, if it does, the amount of solar radi-

© The Author(s), under exclusive license to Springer Nature Singapore Pte Ltd. 2022   55
S. Stevanović, *Overhang Design Methods*,
SpringerBriefs in Architectural Design and Technology,
https://doi.org/10.1007/978-981-19-3012-6_4

ation that enters the interior space through the cell during the given hour. This hourly shading necessity is then taken with a plus or a minus sign in the sum, depending on whether shading of the interior space is beneficial or undesirable during that hour, which is determined by examining whether the indoor thermal comfort exists at that hour without the effects of mechanical systems, artificial lighting and direct solar radiation. The advantage of this method is that it can easily handle curved geometries and that it can work in three dimensions as well by examining a number of parallel shading support surfaces to determine shading necessity patterns for each of their cells.

Implementation of the cellular method in Autodesk Ecotect is described by Kaftan and Marsh in [10]. The amount of solar radiation entering the interior space through a cell during a given hour is calculated by multiplying the amount of hourly direct solar radiation from a weather file with the cosine of the solar incidence angle on the window to be shaded and its glass transmittance, while the indoor thermal comfort is estimated by Ecotect's thermal simulation engine. The shading necessities of cells are then represented visually as a heat map on the shading support surface. One should note here that the cellular method does not prescribe the final overhang shape, but just informs the designer of the necessity to include individual support surface cells in the overhang design to obtain interior thermal comfort, which the designer may then align with other (aesthetical, structural, economical) considerations in order to achieve desired architectural expression.

Sargent et al. [24] base their discussion on the understanding that, since direct solar gains are beneficial for lowering heating loads and undesirable for increasing cooling loads, their desirability at each hour may be expressed through the difference between cooling and heating loads at that hour. More specifically, assume that an annual thermal simulation (with EnergyPlus) is performed for the considered indoor space without any shading device and that $E_{cooling}$ and $E_{heating}$ represent the sensible cooling and heating loads adjusted by energy conversion factors at a given hour. Then the desired fraction of transmitted direct solar energy E_{direct} at that hour may be expressed as

$$T_{desired} = \begin{cases} 1, & E_{cooling} - E_{heating} \leq 0, \\ 1 - \frac{E_{cooling} - E_{heating}}{E_{direct}}, & 0 < E_{cooling} - E_{heating} < E_{direct}, \\ 0, & E_{direct} \leq E_{cooling} - E_{heating}. \end{cases}$$

This type of argument is first used to suggest two ways of defining scholastically symmetric cut-off days for shading, since fixed shading devices block direct solar radiation equally on scholastically symmetric days. In the first case, the cut-off days represent the period with the maximum sum of differences $CDD - HDD$ over all hours in the period, where CDD and HDD represent cooling degree days and heating degree days at the given hour, respectively, while in the second case, the cut-off days represent the period with the maximum sum of differences $E_{cooling} - E_{heating}$ over all hours in the period. (Still, the choice of cut-off times during the cut-off days is left to the designer.)

Sargent et al. [24] further suggest an improvement of Kaftan's cellular method that uses $T_{desired}$ as the estimate of how much direct solar radiation is desirable at a given hour instead of just a yes/no answer about whether shading is undesirable or beneficial during that hour. Their method, called Shaderade, evenly samples points from the window surface and from each point traces the solar rays backward into a user-defined exterior shading support volume, divided into a uniform cell grid, for each hour in the period of overheating. Intersections of each cell and the traced rays are determined in order to quantify parts of $E_{cooling}$, $E_{heating}$ and E_{direct} that the cell is responsible for. Shaderade then determines the cell transmittance T_{cell} as the value between 0 and 1 (in steps of 0.01) that minimises the sum of values $|(E_{cooling} - E_{heating}) - (1 - T_{cell})E_{direct}|$ over all rays that intersect the cell.

A heat map visualising cell transmittances would then help to suggest the shape of a translucent exterior shading. However, for opaque shading devices the cells should have transmittances of either 0 or 1, and a way to quantify their shading impact is to calculate the difference between the part of initial loads ($E_{cooling} - E_{heating}$) that the cell is responsible for and the predicted loads if the cell would have zero transmittance. If this difference, referred to as the *opaque cell inclusion value OCIV* in [24], is positive, then it is beneficial to include the cell in the shading device, while the cell is undesirable if its *OCIV* is negative. The *OCIV* value may then be considered as the importance of the cell for thermal performance, so that the size of the final shading device can be efficiently managed by including only the cells whose *OCIV* values are above a given threshold. Since Shaderade ignores diffuse solar gains and daylighting, one needs to simulate a series of trimmed shadings obtained for different threshold values to fully estimate their behaviour at the end.

The case study was performed in [24] for a south-facing cellular office in three US cities: Anchorage (61.18°N, 150.00°W), Boston (42.37°N, 71.02°W) and Phoenix (33.45°N, 111.98°W). Various conventional shading geometries were compared to three shading devices generated by Shaderade: a horizontal overhang, a perimeter shade and a translucent box. EnergyPlus simulations showed that the Shaderade shadings produced the largest average savings, with a mean heating and cooling load reduction of 26% and a mean carbon emission reduction of 14% across the three climates.

4.2 NURBS Outlined Overhangs

As Stevanović and Stevanović noted in [30], a welcome characteristic of the methods presented in the previous section is that the cells of the shading support surface can be sorted according to computed values which allows one to collect cells in decreasing order of effectiveness in order to obtain an overhang with the required surface area. However, such shapes tend to be serrated, requesting the architect's interference to produce an aesthetic design and attach the overhang to the wall. To accommodate the expected curvilinearity of optimal shading shape, as implied by Kaftan [9], and to ensure smooth overhang design, Stevanović and her coauthors discussed overhangs outlined by NURBS curves in a short series of papers [26, 27, 30, 31].

As explained in [30], NURBS is a widely accepted standard in computer-aided design, engineering and manufacturing for describing and generating smooth curves and surfaces [22]. A NURBS curve is defined by a sequence of control points P_i, $i \in I$ for some index set I, which act as if P_i were connected to the curve by a spring of strength w_i. Each point of the NURBS curve $C(t)$, $0 \le t \le 1$, is actually a convex combination of the control points:

$$C(t) = \frac{\sum_{i \in I} w_i N_i(t) P_i}{\sum_{i \in I} w_i N_i(t)},$$

where $N_i(t)$ are suitably calculated basis functions. The basis functions are determined by a degree d and a knot vector which partitions the interval $[0, 1]$ into knot spans, in such a way that $\sum_{i \in I} N_i(t) = 1$ holds for each $t \in [0, 1]$ and that each basis function has $d + 1$ consecutive knot spans on each of which it reduces to a polynomial of degree d while it is equal to zero outside these knot spans. These conditions ensure that each curve point is determined by $d + 1$ closest control points. Details of computation of basis functions may be found in [22].

The overhang and two side fins for a south-facing window in [30, 31] are placed tightly around the window, in a vertical (fins) or a horizontal plane (overhang) and orthogonal to the wall. Their outer edges are modelled as NURBS curves with the shape of fins controlled by five control points each and the overhang by seven control points. The control points have constant projections to the wall, with only their distance from the wall being changed. The NURBS curves are clamped so they start with the first and end with the last control point, and moreover the overhang shares control points for its two ends with the upper ends of fins, to ensure that shading has continuous look. As EnergyPlus cannot handle NURBS curves directly and can use only triangles and planar quadrangles to model building geometry, calculation of NURBS curve points in preliminary studies [26, 27] was done in EPMacro, the macro language of EnergyPlus. This was a rather cumbersome solution due to the limitations of EPMacro, so a Python package *epnurbs* was developed in [30], which approximates the surfaces of the overhang and the fins with an array of trapezoids in the EnergyPlus model.

In addition to the smoothness of the shading outline, another benefit of using NURBS curves is the ability to control the size of the search space through the number of feasible values that each curve point may take. In [30], each of 15 control points had 9 feasible values, so the search space consisted of 9^{15} curves, while in [31] control points were distributed in 6 groups with control points in the same group taking equal value from the set of 201 feasible values, thus yielding a search space with 201^6 curves. Genetic optimisation, performed through the coupling of jEPlus+EA, epnurbs and EnergyPlus, was looking for appropriate NURBS curves in these search spaces: in [30] it was used to minimise primary energy needed for heating, cooling and lighting a cellular office in the underlying PNNL large office building model [7, 34, 36], which represent realistic construction materials and practices in the US. In [31], genetic optimisation was applied to determine Pareto fronts for heating and cooling energy needed for an apartment room in the underlying PNNL high-rise apartment building model, in both cases for all 16 North American climates for which the PNNL models were developed.

One of the main observations in [30] was that the interior overhang control points in the resulting optimal solutions for a cellular office had relatively similar wall distances, implying that the interior part of the overhang's outline was almost a straight line and thus partially rejecting the expectation that optimal shading shapes should be curvilinear. Moreover, it turned out that with an increase of primary energy for at most 0.24%, the shading structure could be significantly simplified by identifying wall distances of successive control points, which formed six independent shading regions: the lower and the upper part of the western fin (two control points each), joint of the western fin and the overhang (one control point), interior part of the overhang (five control points), joint of the eastern fin and the overhang (one control point) and the rest of the eastern fin (four control points). This finding was confirmed in [31] for the overhang and side fins of an apartment room window, where Pareto fronts of two genetic optimisation runs, one for shadings whose control points wall distances are set independently and another for shading whose control points are grouped as above, were shown to have extremely similar shapes with small root mean square distance between them. Thus, such grouping of control points gives a simple and natural division of shading into a small number of basic constituents that have the most impact on its performance, and which with a smaller number of defining parameters can be used in further studies.

As heating and cooling loads were the only objective functions for shadings of apartment room window in [31], further discussion was focused on the convex hull of the Pareto front for shadings with grouped control points that, as explained in Sect. 3.3.1, contains shadings that minimise any linear combination of heating and cooling loads. In particular, the primary energy needed for heating and cooling is proportional to a particular linear combination obtained for the ratio r, expressed in terms of source energy conversion factors and district heating and cooling efficiencies, which represents the relative importance of the cooling load with respect to the heating load. Each climate has its own threshold value so that if r is below the threshold then window shading becomes undesirable, as the reduction of cooling load it yields during summer is overcome by the increase in heating load during winter. On the other hand, if r is sufficiently large then the reduction of cooling load becomes so important that optimal shadings get increasing depths, almost always to the full extent allowed. Figure 4.1 shows optimal convex hull shadings for the value $r = 0.3038$ that corresponds to current values of source energy conversion factors and district heating and cooling efficiencies reported in [4].

Visualisations of wall distances of control point groups in optimal convex hull shadings, as functions with respect to varying values of the ratio r, showed that most of their graphs have a sigmoid-like shape: once each wall distance becomes positive, it increases almost linearly with the increase in r, until it reaches its maximum value. It was found that the arctangent function provided satisfactory fitting of these functions in most cases, with issues arising only when wall distances of control point groups

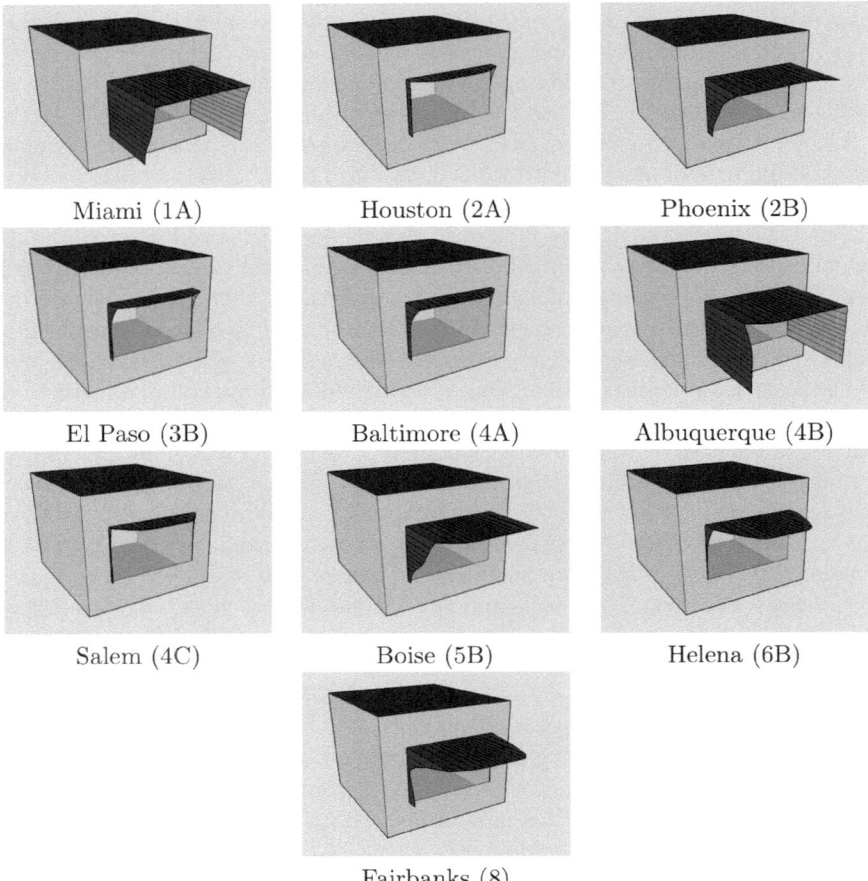

Miami (1A)   Houston (2A)   Phoenix (2B)

El Paso (3B)   Baltimore (4A)   Albuquerque (4B)

Salem (4C)   Boise (5B)   Helena (6B)

Fairbanks (8)

Fig. 4.1 Sketchup visualisations of convex hull shadings for a window of an apartment room with minimal primary energy for the ratio $r = 0.3038$ from [31]. In the remaining six considered locations, Memphis, San Francisco, Chicago, Vancouver, Burlington and Duluth, minimal primary energy is attained in the base case of the model without any shading. Three shading results require particular attention. First, the pronounced overhang for Fairbanks actually suggests that some amount of shading is required for the considered model at this location as well. However, due to its northern location, the overhang has to be extended further outside to prevent midday solar heat gain in summer, so a vertically placed shading device or a better performing solar control glazing are better alternatives for this location. On the other hand, extremely deep shadings for Miami and Albuquerque suggest that genetic optimisation was trying to prevent not only direct but also diffuse radiation from reaching the window, as it apparently has a large impact on cooling load in these locations. In these cases, a smaller window-to-wall ratio is another alternative in addition to a vertically placed shading device or a better performing glazing

experience sudden jumps, thus greatly increasing the slope of the appropriate graph, which happened for 3 out of 16 locations where successive convex hull shadings were found far apart in the Pareto front. For the remaining locations, the sequence of shadings with fitted wall distances of control point groups for varying values of r produced a satisfactory approximation of heating and cooling loads of Pareto front shadings, indicating that sigmoid functions might be used in further studies to obtain theoretical models that would be able to approximate Pareto front shadings for other locations as well.

4.3 Trimming Windows to Fit the Overhang

Motivated by the observation that lower corners of windows are hard to protect by overhangs, Sherif and El Zafarany [25] reverse the usual problem of designing the shading device to fit a window and instead suggest designing the window to fit a shading device. Methods presented in earlier sections may easily lead to excessive shading devices that obstruct the view outside and reduce diffuse radiation needed for daylighting, and a simple act of cutting the window corners may significantly reduce the size of necessary shading devices.

In their approach, Sherif and El Zafarany fix the shading device and identify the pattern of shadows cast by the device on the surface of the wall, where a window should be located, during the period of overheating. The wall surface is divided into a two-dimensional array of cells, and for each cell the energy, that would be added to cooling loads if the window would occupy that cell, is obtained by summing for each time step the intensity of direct solar radiation reaching the cell multiplied by the cosine of the angle of incidence and transmittance of glass at this angle, together with the amount of diffuse solar radiation multiplied by the sky view factor. Visualising the relative values of so-obtained cell transmitted energies on the wall surface clearly shows more shaded cells with lower transmitted energies that are fit to be a part of the window. Defining different threshold values for cell acceptance based on their transmitted energy would yield a series of possible window shapes, which can then be subjected to further design considerations.

As a case study, window designs were proposed for a horizontal overhang on a south-facing façade of a building in a cooling-dominated climate of Cairo, Egypt (30°N, 31°E), with shading required from February 21 to October 21. The proposed designs that fit a selected pattern of cell transmitted energies are shown in Fig. 4.2.

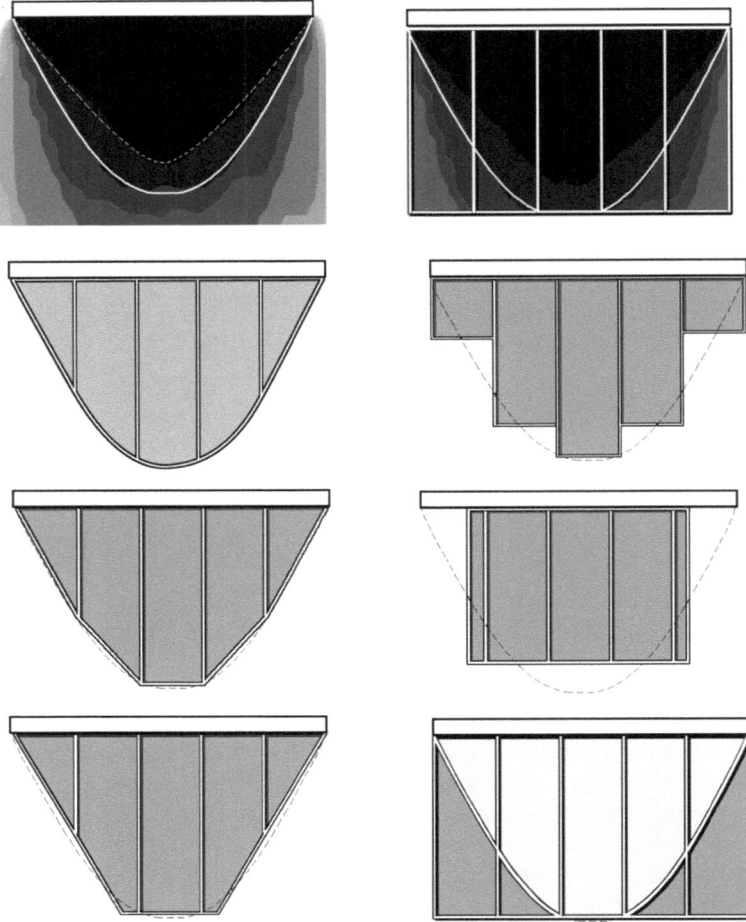

Fig. 4.2 Possible window design configurations fitting the horizontal overhang [25]

4.4   Shading Retrofits

We discuss here a few recent studies that consider the suitability of overhangs as exterior shading devices in building retrofits.

Huang et al. [8] analysed a case study of exterior shading retrofit project for a university campus in Hong Kong. While solar film and slat-type sunshade were considered as well, offering much higher efficiency in reducing solar heat gains, overhang shading was used in the actual retrofit (see Fig. 1.1) due to its structural reliability. Hong Kong experiences a number of typhoons, tropical storms and thunderstorms annually, so safety factors under extreme weather conditions require exceptionally strongly built structures. Measurements after the overhangs were installed showed that the annual cooling load of affected interior spaces was reduced by 44.1%, with the overhangs on west- and east-facing windows having much more effect than the overhangs on south-facing windows, in line with Cheung and Chung's observations [2] that the azimuth angle ranging from $-105°$ to $-85°$ and from $+85°$ to $105°$ are major contributors to expected sunlight duration in summer (see Sect. 2.2). Economic feasibility was assessed through the life cycle environmental impact of installed overhangs. The use of high-strength building materials required large energy expenditures and CO_2 emissions (the overhangs weigh approximately 240 kg per m^2 of windows), so the overhangs make a little contribution to energy conservation and even have a negative effect on CO_2 emissions during their life cycle. Actually, it is estimated that such a shading system saves only 53,400 HKD annually, while its installation cost almost 30 million HKD, leading to the conclusion that overhangs may not necessarily be a useful retrofit option in low latitude subtropical climate areas such as Hong Kong.

Cho et al. [3] analysed feasible exterior shading devices for installation in a high-rise residential building with a curtain-wall façade in Seoul, Korea. Thermal and visual performance was initially considered for five types of shading devices, with three alternative versions of each type, and for three façade orientations (eastern, western and southern) from May to September. Shading factor and direct solar heat gains were computed with DOE-2.1E, while daylight factor, two variants of daylight autonomy with minimum required illuminances of 300 lux and 500 lux, respectively, and useful daylight illuminance from 8 am to 6 pm were computed with DAYSIM. Considered shading devices did not reduce daylighting performance significantly, as their depth was limited to 0.6 m, while the window module had dimensions 4.0 m × 3.2 m. This initial analysis recommended a horizontal overhang with a depth of 0.6 m and a vertical panel with 0.6m height in front and above the window module, shown in Fig. 4.3, for actual installation on a building.

The horizontal overhang and the vertical panel were further used in building performance simulations with HVAC modelled in detail, for four apartments of size 150 m^2 each, having three different orientations (south, south-west and south-east). The overhang reduced direct solar heat gains by 21–27%, while the panel reduced them by 18–20%. Cooling energy costs and costs of implementing shading devices, which include costs for principal and accessory materials, for manufacturing, assembly, shipping and installation, were used in economic analysis that showed that the minimum simple payback period for the overhang is 3.4 years, while it is 8.7 years for the vertical panel.

Finally, full-scale models of the overhang and the panel were produced and subjected to wind pressure testing and vibration analysis, as noise-related issues are the biggest complaints among apartment dwellers in Korea. Both devices turned out to be acceptable, with the horizontal overhang less susceptible to vibrations than the vertical panel, leading to the final recommendation for their installation in high-rise residential buildings in Korea.

Nikolić et al. [19] consider retrofitting the typical detached two-storey house built in Belgrade, Serbia, between the 1960s and 1990s, with overhanging shading in the

Fig. 4.3 Shading devices considered for installation on a high-rise residential building: the horizontal overhang (top) and the vertical panel (bottom) [3]

form of either roof extensions that shade the first floor or addition of balconies to the first floor which then shade the ground floor. Such overhangs are added to all four façades with the depths varying between 0.2 and 3 m. Heating, cooling and lighting demands were simulated in EnergyPlus, while the minimisation of the annual primary energy for heating, cooling and lighting, together with the energy needed to build the actual overhangs was performed in GenOpt with the Hooke–Jeeves algorithm. It was observed that one can achieve annual energy saving up to 3.69% when only roof overhangs are installed, and up to 7.12% when both roof overhangs and balconies are added. However, while the optimal southern overhangs have depths of 0.95 m for roof overhangs and 0.7 m for balconies, the optimal eastern and western overhangs turn out to be oversized with depths between 1.9 and 2.6 m, due to the small incidence angle of solar rays in the morning and the afternoon during the summer months.

4.5 PV Integrated and Movable Overhangs

While there are quite a few implementations of PV cells within shading devices in real buildings nowadays (see [38] for a list of examples), only Yoo et al. [37] seem to report performance measurements for their implementation within overhangs. The overhangs in question are made from semitransparent PV modules placed above south-facing windows and at a minimum distance from the walls of the building of Samsung Institute of Engineering and Construction Technology in South Korea. Semitransparent modules admit some daylight in summer, while the distance from the wall improves ventilation for cooling of modules. The modules were installed at a fixed tilt angle for safety and costs, while aesthetic reasons forced installation at a higher tilt angle (55.5°) than the optimum one (32°).

Mandalaki et al. [18] studied effects of integrating PV cells in exterior shading devices by investigating the balance between heating and cooling energy needs (simulated with EnergyPlus), artificial lighting energy (simulated with Radiance) and electricity generated by PV cells (simulated with Ecotect). Thirteen types of shading devices, designed to block direct solar radiation from June to mid-September for a cellular office located in Athens and Chania (Crete) in Greece, were examined and a physical model in 1:10 scale was built and tested in real sky conditions. By using mono-crystalline cells with 15% efficiency, it was shown that PV cells integrated into each of the considered south-facing shading devices can generate more electricity than what is needed for lighting. A combination of overhang and triangular fins is among the most efficient of these devices, both in terms of the lowest primary energy needed for heating, cooling and lighting and in terms of generated electricity relative to the area that PV cells occupy.

A short series of papers [32, 33, 39] discusses optimal tilt angles for installation of PV integrated overhangs in Hong Kong. Sun and Yang [33] developed a theoretical model for annual electricity generation of PV integrated overhangs and cooling loads of associated windows based on the Perez sky model [21], the simplified model of Lu and Yang [17] for the maximum power output of PV modules, the Powell–Yellott model [23] for solar heat gain factor and the radiant time series method [1] for cooling load calculations. Analysis of ten tilt angles ranging uniformly from 0° to 90° showed that, taking into account both electricity generation and cooling loads, the optimum tilt angle for PV integrated overhang, whose vertical projection covers the wall area between windows on successive storeys, is between 30° and 50°, while the horizontal position is optimal for a PV integrated overhang of fixed depth.

Sun et al. [32] improved the previous theoretical model for annual electricity generation and cooling loads to analyse optimum tilt angles for PV integrated overhangs with five different orientations: east, south-east, south, south-west and west. Overhangs had the maximum horizontal projection of 1.5 m, while the vertical projection was given in percentage of the wall area between windows on successive storeys. Similar to the previous study, optimum tilt angles varied from 30° to 50° depending on the orientation and the wall utilisation percentage, with the south-west-facing overhangs offering higher annual electricity savings per unit area of PV cells than other orientations.

Zhang et al. [39] turned to EnergyPlus and its *EquivalentOneDiodeModel* to simulate power output of PV integrated overhangs. This is the four-parameter equivalent circuit model, developed largely by Townsend [35] and detailed in Duffie and Beckman [5], which was first incorporated as a component for TRNSYS back in the 1990s [6]. Ten tilt angles (0°–90° with 10° steps) and five wall orientations (east, south-east, south, south-west, west) were considered, simulating PV electricity generation and heating, cooling and artificial lighting loads of the associated cellular office. Accuracy of the EquivalentOneDiodeModel was verified through comparisons with experimental data measured over one month (January 2015) at an outdoor test bed with PV cells installed at the tilt angle of 55°, obtaining a relative error of 5%. Unlike the result of Sun and Yang [33], this study obtained that the optimal tilt angle to install a PV integrated overhang of fixed depth at the southern façade in Hong Kong is 20°, with smaller tilt angles (0°–40°) achieving higher electricity generation. As a matter of fact, as the electricity generation peaks at tilt angles of 20° and 30°, which enclose the location latitude of 22.3 °N, the study could have further considered a finer resolution of tilt angles between these two values to obtain a more precise estimate of the optimal tilt angle and check its correlation with the location latitude. Another aspect that could have been added to these studies is an estimate of economic feasibility of installing PV integrated overhangs at the optimal tilt angles, which is a common research question in solar technology studies [28, 29].

Moving away from Hong Kong studies, Li et al. [16] also used the *Equivalent-OneDiodeModel* from EnergyPlus to determine annual and monthly optimum tilt angles for installation of PV integrated overhangs for five cities: Harbin, Beijing, Changsha, Kunming and Guangzhou in different climatic regions of China. Instead of an office setting, simulations were performed for a residential apartment, but with a bit unusual requirement (for residential spaces) of minimum illumination level of 400 lux in the centre of each room at desk height, calculating PV electricity generation and heating, cooling and artificial lighting loads. A common thread in this and previous studies is the unimodal dependence of annual electricity generation on tilt angles, which first increases with the tilt angle from 0° to the optimum value and then decreases towards 90°. While the results naturally show that the use of monthly optimum tilt angles would be superior to annual optimum tilt angles, fixed installation at annual optimum tilt angles is still recommended due to safety and associated costs.

Paydar [20] elaborated on the idea of using PV panels as a movable overhang, based on the observed compatibility between the PV electricity generation and the overhang shading efficacy. First, keeping the PV panel surface nearly orthogonal to the solar rays throughout the year will lead to higher electricity generation. In summer, when the PV panel should be set at a smaller tilt angle due to the high solar altitude, the panel should be positioned right over the window to block direct solar radiation and decrease cooling loads. In winter, when the PV panel should be set at a higher tilt angle due to the low solar altitude, the panel should be positioned above the window to allow the entry of direct solar radiation and decrease heating loads. Paydar's proposal for the construction of such a movable PV integrated overhang is illustrated in Fig. 4.4.

The construction was optimised for use as overhangs above two southern windows in an apartment located in Tehran, Iran. Using EnergyPlus as the simulation engine for computing both the heating and cooling loads and the PV electricity generation, the thermal impact of the overhang depths ranging from 0 to 1 m in 0.1 m steps was simulated monthly to observe that the optimal overhang depth should be 1m from

Fig. 4.4 Proposed construction of a movable PV integrated overhang from [20]

May to October and 0m from November to March (the optimum overhang depth in April is 0.1 m, but the thermal impact of the 0 m depth is within 2% to the minimum value). For separate simulation of PV electricity generation for tilt angles ranging from 0° to 90° with 10° steps, the optimum tilt angle varied between 20° and 60° in different months.

As the optimum depths and tilt angles do not combine well when the PV panel has a fixed position, Paydar opted for this new construction. The construction was then optimised according to the length of the retaining arms and the vertical position at which the arms are connected to the wall, while the depth of the PV panel was fixed at 1 m. The optimum monthly angles were selected based on the difference between the generated PV electricity and the electricity consumed by the air-conditioning system. While there is a total of six different optimum tilt angles throughout the year, the optimum angle is constant from May to September and equal to 25°, and then again constant from December to March and equal to 65°. With the aim of reducing the number of manual updates, it is shown that by setting the PV panel at a 25° tilt angle from April to October, and then at 65° tilt angle from November to March decreased the system efficiency by only 3%. The construction operated in this way turns out to be more effective than the cases when the PV panel is installed at a fixed tilt angle either above the window to act as an overhang or at the roof.

In a recent series of papers [12–15], Krarti studied the performance of dynamic overhangs that may rotate around a horizontal axis fixed above the window or slide away horizontally from the window, and that may also be PV integrated (see Fig. 4.5). In this way, besides being able to select the optimal tilt angle for shading and/or generating PV electricity, one can also move the overhang off the window during winter periods when its shading effects are undesirable. The aim of the studies is

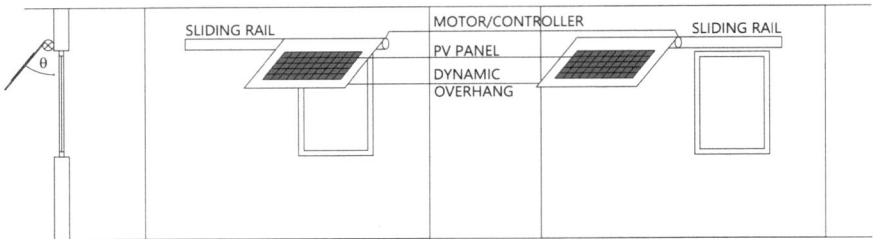

Fig. 4.5 Basic design of the sliding–rotating overhang placed above a window, with or without integrated PV

to assess the potential of such dynamic overhangs for reducing annual net energy demand in residential buildings, i.e., for reducing heating and cooling energy use while increasing generated PV electricity, if PV integrated. The overhang of depth 0.6m is running along the width of the window, and the building model is either a ranch house with windows in all four directions or a residential room with a southern facing window located in the representative US climates of Phoenix, San Francisco, Boulder and either Chicago (in [12, 13, 15]) or Minneapolis (in [14]).

The operation schedules for the overhangs were considered on monthly, daily and hourly scales. The technique of distinct building energy models was used to simulate their behaviour on monthly scales, so the energy performance of dynamic overhangs is actually determined through the monthly simulation of static overhangs set at various angle positions in a number of copies of the building model. On the other hand, hourly operating scenarios were simulated using parametric behaviour maps [11], which define solar transmittance fractions specific to exterior shading of the window in DOE-2 simulation engine for the representative days in an annual analysis. Daily operation schedules were simulated with distinct building energy modelling in [12, 13, 15] and with parametric behaviour maps in [14]. In cases when PV arrays are integrated in overhangs, the SAM tool was used to estimate the shading effects of the house and its surrounding on the PV arrays and the generated PV electricity.

While the tilt angle of the rotating overhang could in principle be set to any value θ between given lower and upper limits, the studies considered only a discrete set of possible tilt angles to simplify the analysis, $0°$, $45°$, $90°$, $135°$ and $180°$, as well as the location latitude and latitude $\pm15°$ for PV integrated overhangs. Three control scenarios were considered for selecting the overhang tilt angle within any given timeframe. In cases without PV integration, the tilt angle was adjusted to reduce annual heating and cooling energy consumption. With PV arrays installed on the top side of the overhang, the tilt angle was set either according to load-tracking, which minimises heating and cooling energy uses within that timeframe, or according to PV-tracking, which minimises the difference between the combined heating and cooling energy demand and the generated PV electricity. To reduce computational efforts, optimal tilt angles are determined for one representative week for each month when daily schedules are considered, and from sunrise to sunset for one representative day of each month when hourly adjustments are used.

Without PV arrays, horizontal static overhangs are beneficial in the warmer climates of Phoenix and San Francisco (28.6% and 3.4% of energy savings compared to the case without overhangs, respectively), while they actually increase the annual HVAC energy use in colder climates of Boulder and Chicago [12]. On the other hand, the dynamic overhangs always perform better than static overhangs and no-overhang options for all considered climates, especially in cases of integrated PV arrays. For example, for a house located in San Francisco, the PV integrated sliding–rotating overhang achieves 57.8% annual energy savings relative to the no-overhang case, while the PV integrated static overhang achieves 44.2% savings [14]. For a room with 30% WWR and low-e glazing in the warm climate of Phoenix, the PV integrated rotating overhang can save up to 90% in annual net energy demand compared to the no-overhang case and up to 75% compared to the PV integrated static horizontal overhang [13]. In cold climates, reductions in net energy savings are smaller but still substantial: in Boulder, the PV integrated rotating overhang saves up to 35% in net energy demand compared to the no-overhang option and up to 15% compared to the PV integrated static horizontal overhang [13].

As expected, the highest energy saving potential is achievable by hourly controls, followed by daily controls, and then by monthly controls. This difference is more pronounced for dynamic overhangs without PV arrays: for example, if the overhang tilt angle is allowed to vary between 45° and 135°, the hourly update schedule achieves 42% savings in combined heating and cooling energy use for a house located in San Francisco, significantly higher than 27% savings for the daily updates and 24% for the monthly updates [12]. However, for PV integrated dynamic overhangs most of the energy-saving potential turns out to be captured already by the monthly operating scenarios, due to the slowly changing nature of solar radiation which determines both the overhang shading effects and the generated PV electricity.

Two separate behaviours are observed concerning the optimal tilt angles. In certain setups, the dynamic overhangs are mostly set at extreme tilt angles. For example, without PV arrays the rotating overhang in the colder climates of Chicago and Boulder was mostly set at the upper limit to minimise its shading effects during winter and colder months, while in the hot climate of Phoenix it was mostly set at the lower limit to reduce solar heat gains [12]. A similar effect was observed for load-tracking controls of PV integrated rotating overhangs which aim to maximise solar gains in winter and minimise them in summer [13]. In other cases, the optimal tilt angles are more balanced. For example, the optimal tilt angle for a rotating overhang without PV arrays in the mild climate of San Francisco is well dispersed over all angles in the allowed range, with a slight preference for the 90° tilt angle [12]. Similarly, PV-tracking controls for PV integrated rotating and sliding–rotating overhangs aim to offset the heating and cooling loads by the generated PV electricity, and so the optimal tilt angle varies on a monthly basis [13, 14]. Nevertheless, it is observed that the PV integrated sliding overhang with the tilt angle fixed at the location latitude (so that it can only slide over the window or away from the window) performs very close to the best achievable dynamic overhang combinations, recommending its choice for implementation as it simplifies overhang design and reduces maintenance costs [15].

The actual performance of dynamic overhangs is found to be very much dependent on location climate, windows-to-wall ratio and overhang depth. With integrated PV, larger window width and/or larger overhang depth provide more area for installation of PV modules on the overhang surface, which reduce net energy demand by generating more PV electricity. In particular, when the overhang depth doubles from 0.6 to 1.2 m, the PV integrated sliding–rotating overhang with tilt angles limited between 45° and 90° increases annual net energy demand savings from 16.8% for 0.6 m depth to 31.3% for 1.2 m depth, compared to the no-overhang case for a house located in Boulder [14]. Even without integrated PV arrays, sliding–rotating overhangs decrease cooling demand in summer with increasing overhang depth, while having equal no-shading benefits when slid off the window during winter, thus effectively increasing net energy savings with increasing overhang depth, unlike the static overhangs [14].

The PV integrated dynamic overhangs perform best in mild and hot climates, significantly reducing the net energy demand and even reaching the net-zero energy goal. For example, the PV integrated sliding overhang with tilt angle fixed at latitude location can reduce heating and cooling demand for the ranch house with low-e glazing and 30% WWR by 100% in San Francisco and 92.3% in Phoenix, more than double those achieved in Boulder (57.1%) and Chicago (39.2%) [15]. Such overhangs then present a worthwhile alternative to rooftop PV systems. As a matter of fact, energy savings from dynamic overhangs are even higher than those reported for exterior automated Venetian blinds [14].

It remains to be seen how PV integrated dynamic overhangs may perform in office buildings, where the windows-to-wall ratio may be too large to leave enough free wall space to slide the overhang horizontally off the window. Depending on the building design, the overhang could be perhaps slid off vertically to the bottom of the window when its shading effects are not needed, provided it does not shade the window at the floor below in such a case. One also needs to study the influence of dynamic overhangs on daylighting, a question that was not considered in [12–15].

References

1. ASHRAE (2005) Handbook: fundamentals. ASHRAE, Atlanta
2. Cheung HD, Chung TM (2007) Analyzing sunlight duration and optimum shading using a sky map. Build Environ 42:3138–3148
3. Cho J, Yoo C, Kim Y (2014) Viability of exterior shading devices for high-rise residential buildings: case study for cooling energy saving and economic feasibility analysis. Energ Buildings 82:771–785
4. Deru M, Torcellini P (2007) Source energy and emission factors for energy use in buildings. Technical Report NREL/TP-550-38617. National Renewable Energy Laboratory, Golden
5. Duffie JA, Beckman WA (1991) Solar engineering of thermal processes. Wiley, New York
6. Eckstein JH (1990) Detailed modeling of photovoltaic components. MS thesis. Solar Energy Laboratory of the University of Wisconsin–Madison

7. Goel S, Athalye R, Wang W, Zhang J, Rosenberg M, Xie Y, Hart R, Mendon V (2014) Enhancements to ASHRAE Standard 90.1 prototype building models. Technical Report PNNL-23269. Pacific Northwest National Laboratory, Richland
8. Huang Y, Niu Jl, Chung Tm (2012) Energy and carbon emission payback analysis for energy-efficient retrofitting in buildings—Overhang shading option. Energ Build 44:94–103
9. Kaftan E (2001) The cellular method to design energy efficient shading form to accommodate the dynamic characteristics of climate. In: Pereira FOR, Rüther R, Souza RVG, Afonso S, Neto JABdC (eds) Proceedings of PLEA 2001: the 18th international conference on passive and low energy architecture, Florianopolis, Brazil, 2001 Nov 7–9. PLEA, pp 829–833
10. Kaftan E, Marsh A (2005) Integrating the cellular method for shading design with a thermal simulation. In: Santamouris M (ed) Proceedings of the international conference "Passive and low energy cooling for the built environment", Santorini, Greece, 2005 May 19–21. Heliotopos Conferences, Santorini, pp 965–970
11. Kim H, Clayton MJ (2020) Parametric behavior maps: a method for evaluating the energy performance of climate-adaptive building envelopes. Energ Build 219:110020
12. Krarti M (2021) Evaluation of energy performance of dynamic overhang systems for US residential buildings. Energ Build 234:110699
13. Krarti M (2021) Impact of PV integrated rotating overhangs for US residential buildings. Renew Energ 174:835–849
14. Krarti M (2021) Performance of PV integrated dynamic overhangs applied to US homes. Energy 230:120843
15. Krarti M (2021) Evaluation of PV integrated sliding-rotating overhangs for US apartment buildings. Appl Energ 293:116942
16. Li X, Peng JQ, Li NP, Wang M, Wang CL (2017) Study on optimum tilt angles of photovoltaic shading systems in different climatic regions of China. Procedia Eng 205:1157–1164
17. Lu L, Yang HX (2004) A study on simulations of the power output and practical models for building integrated photovoltaic systems. J Sol Energ-T ASME 126:929–935
18. Mandalaki M, Zervas K, Tsoutsos T, Vazakas A (2012) Assessment of fixed shading devices with integrated PV for efficient energy use. Sol Energy 86:2561–2575
19. Nikolić D, Djordjević S, Skerlić J, Radulović J (2020) Energy analyses of serbian buildings with horizontal overhangs: a case study. Energies 13:4577
20. Paydar MA (2020) Optimum design of building integrated PV module as a movable shading device. Sustain Cities Soc 62:102368
21. Perez R, Ineichen P, Seals , Michalsky J, Stewart R (1990) Modelling daylight availability and irradiance components from direct and global irradiance. Sol Energy 44:271–289
22. Piegl L, Tiller W (1997) The NURBS book. Springer, Berlin
23. Powell GL, Yellott JI (1980) Solar heat gain factors on average days. Proc. Annu. Meet. – Am. Sect. Int. Sol. Energy Soc. (United States) 3(2):826–830
24. Sargent JA, Niemasz J, Reinhart CF (2011) Shaderade: combining Rhinoceros and Energyplus for the design of static exterior shading devices. In: Proceedings of building simulation 2011: the 12th international IBPSA conference, Sydney, Australia, 2011 Nov 14–16. IBPSA, pp 310–317
25. Sherif A, Zafarany AE (2011) Designing the window to fit a shading device: a reversed method for optimizing energy efficient fenestration. In: Leclercq P, Heylighen A, Martin G (eds) Proceedings of CAAD futures 2011: designing together, Liège, Belgium, 2011 Jul 4–8. Les Éditions de l'Université de Liège, Liège, pp 383–399
26. Stevanović S (2016) Search for the optimal shape of fixed exterior shading in a cooling-dominated climate. In: Kitanovski A, Poredoš A (eds) Proceedings of ECOS 2016: the 29th international conference on efficiency, cost, optimization, simulation and environmental impact of energy systems, Portorož, Slovenia, 2016 Jun 19–23. Faculty of Mechanical Engineering, University of Ljubljana, Ljubljana
27. Stevanović S, Naumoska VE, Ralev M (2016) Assessment of optimal shape of exterior shading for offices in hot summer continental climate. In: Tepavčević B, Stojaković V (eds) Proceedings

of the 4th international regional eCAADe workshop "Between computational models and performative capacities", Novi Sad, Serbia, 2016 May 19–20. eCAADe and Faculty of Technical Sciences, University of Novi Sad, Novi Sad, pp 10–23

28. Stevanović S, Pucar M (2012) Investment appraisal of a small, grid-connected photovoltaic plant under the Serbian feed-in tariff framework. Renew Susta Energ Rev 16:1673–1682

29. Stevanović S, Pucar M (2012) Financial measures Serbia should offer for solar hot water systems. Energ Build 54:519–526

30. Stevanović S, Stevanović D (2018) Optimisation of curvilinear external shading of windows in cellular offices. PLoS ONE 13:e0203575

31. Stevanović S, Stevanović D, Dehmer M (2019) On optimal and near-optimal shapes of external shading in apartment buildings. PLoS ONE 14:e0212710

32. Sun LL, Lu L, Yang HX (2012) Optimum design of shading-type building-integrated photovoltaic claddings with different surface azimuth angles. Appl Energ 90:233–240

33. Sun LL, Yang HX (2010) Impacts of the shading-type building-integrated photovoltaic claddings on electricity generation and cooling load component through shaded windows. Energ Build 42:455–460

34. Thornton BA, Wang W, Cho H, Xie Y, Mendon VV, Richman EE et al (2011) Achieving 30% goal: energy and cost saving analysis of ASHRAE/IES standard 90.1-2010. Technical Report PNNL-20405. Pacific Northwest National Laboratory, Richland

35. Townsend TU (1989) A method for estimating the long term performance of direct-coupled photovoltaic systems. MS thesis. Solar Energy Laboratory of the University of Wisconsin–Madison, Madison

36. United States Department of Energy. Building energy codes program: commercial prototype building models. https://energycodes.gov/prototype-building-models. Accessed Feb 2022

37. Yoo SH, Lee ET, Lee JK (1998) Building integrated photovoltaics: a Korean case study. Sol Energy 64:151–161

38. Zhang X, Lau SK, Lau SSY, Zhao Y (2018) Photovoltaic integrated shading devices (PVSDs): a review. Sol Energy 170:947–968

39. Zhang W, Lu L, Peng JQ (2017) Evaluation of potential benefits of solar photovoltaic shadings in Hong Kong. Energy 137:1152–1158

Appendix
Locations and Climates of Overhang Case Studies

Abstract This appendix contains the map and the list of locations of overhang case studies from the research literature classified according to the major climate types and location latitude.

Keywords Overhang case studies · Köppen–Geiger climate classification · Map · Locations

As we have seen from the main text, thermal and daylighting performance of overhangs depends on a number of factors, among which the amount of available solar radiation, temperature fluctuations and sky conditions play a great role. In their overview of solar shading systems for buildings, Bellia et al. [7] already pointed out the difficulty of comparing shading performance results from different studies due to large differences in building characteristics and performance indices. Accordingly, only basic data about the overhang case studies is given in the following table: their location with longitude and latitude, the Köppen–Geiger climate type and the literature reference. The table is sorted by the climate type first, and then by the location latitude, as these two values determine a good part of the environmental impact on the building. Hence, this table serves as a form of an index to published overhang case studies, which enables one to quickly find case studies performed for locations with similar climates or in the vicinity of the chosen location. The data is accompanied by the map of case study locations shown in Fig. A.1, which is drawn upon the underlying world climate map from [34].

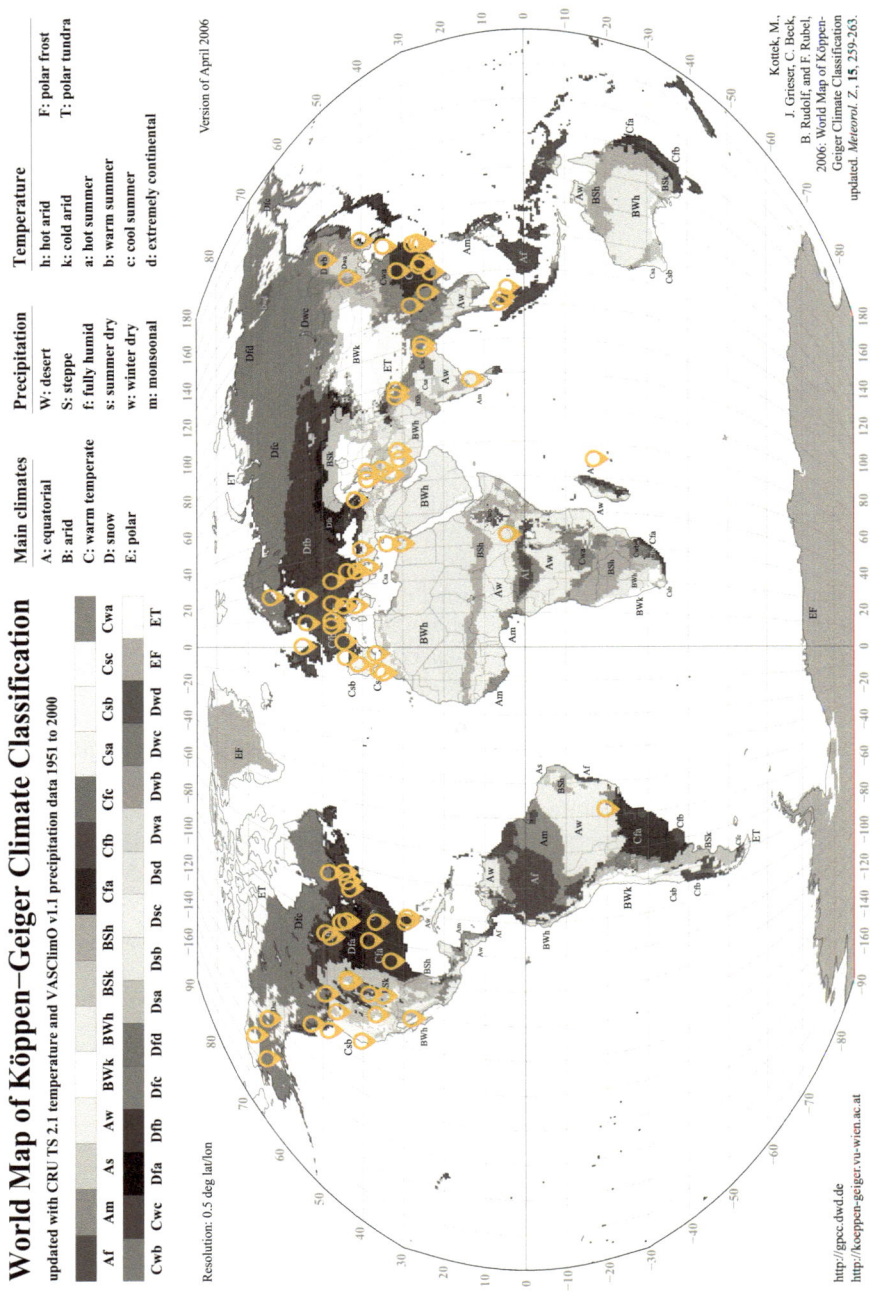

Fig. A.1 The map of case study locations and climate distribution (the underlying map from [34])

Legend of the Köppen–Geiger climate classification system: *Af* = tropical rainforest climate; *Am* = tropical monsoon climate; *Aw* = tropical savanna climate; *BWh* = hot desert climate; *BWk* = cold desert climate; *BSh* = hot semi-arid climate; *BSk* = cold semi-arid climate; *Csa* = Mediterranean hot summer climate; *Csb* = Mediterranean warm summer climate; *Cwa* = subtropical humid monsoon climate; *Cfa* = subtropical humid climate; *Cfb* = temperate oceanic climate; *Dwa* = continental humid hot summer monsoon climate; *Dfa* = continental humid hot summer climate; *Dfb* = continental humid warm summer climate; *Dfc* = subarctic climate; *Dsb* = continental humid warm summer Mediterranean-influenced climate; *Dsc* = subarctic Mediterranean-influenced climate.

Climate	Latitude (°)	Longitude (°)	City	Country	Refs.
Af	0.32	32.58	Kampala	Uganda	[26]
Af	1.35	103.82	Singapore	Singapore	[74]
Af	3.14	101.69	Kuala Lumpur	Malaysia	[39]
Af	5.41	100.30	Penang	Malaysia	[20]
Am	19.23	110.47	Qionghai	China	[75]
Aw	9.93	78.12	Madurai	India	[69]
Aw	−20.88	55.52	Reunion Island	France	[15, 40]
Aw	22.57	88.36	Kolkata	India	[23]
Aw	23.00	120.23	Tainan	Taiwan	[12]
Aw	23.16	89.22	Jessore	Bangladesh	[2]
Aw	25.90	−80.28	Miami	USA	[58, 67, 68]
Aw	26.14	−81.79	Naples (FL)	USA	[23]
BWh	24.14	−110.31	La Paz	Mexico	[24]
BWh	27.12	56.22	Bandar Abass	Iran	[18]
BWh	27.18	31.19	New Assiut City	Egypt	[3]
BWh	29.10	58.35	Bam	Iran	[72]
BWh	30.04	31.24	Cairo	Egypt	[4, 63, 64]
BWh	33.45	−111.98	Phoenix	USA	[35–38, 60, 65, 67, 68]
BWk	31.77	−106.48	El Paso	USA	[67, 68]
BWk	31.87	54.27	Yazd	Iran	[18]
BSh	28.38	75.61	Pilani	India	[25]
BSh	28.88	77.35	New Delhi	India	[22, 31]
BSh	29.60	52.53	Shiraz	Iran	[56]
BSh	31.63	−7.98	Marrakech	Morocco	[29, 61]
BSk	33.57	−7.59	Casablanca	Morocco	[61]
BSk	34.68	−1.90	Oujda	Morocco	[61]
BSk	35.08	−106.65	Albuquerque	USA	[67, 68]
BSk	35.69	51.39	Tehran	Iran	[17, 53, 59, 62]
BSk	38.03	46.17	Tabriz	Iran	[18]
BSk	39.90	116.41	Beijing	China	[27, 42]
BSk	40.02	−105.27	Boulder	USA	[35–38, 43]
BSk	43.62	−116.20	Boise	USA	[67, 68]
BSk	46.59	−112.04	Helena	USA	[67, 68]
Csa	35.51	24.02	Chania	Greece	[45]
Csa	36.55	53.00	Sari	Iran	[18]
Csa	36.87	30.73	Antalya	Turkey	[33]
Csa	37.39	−5.98	Seville	Spain	[19, 41, 44]
Csa	37.98	23.72	Athens	Greece	[45, 50, 54]
Csa	38.12	13.36	Palermo	Italy	[6]
Csa	40.42	−3.70	Madrid	Spain	[32]
Csa	40.64	22.94	Thessaloniki	Greece	[5]
Csa	41.39	2.17	Barcelona	Spain	[1, 54]
Csa	41.78	12.49	Rome	Italy	[6, 46]

(continued)

Csb	37.77	−122.42	San Francisco	USA	[35–38, 67, 68]
Csb	44.94	−123.04	Salem	USA	[67, 68]
Cwa	21.03	105.83	Hanoi	Vietnam	[23]
Cwa	22.00	120.74	Hengchuen	Taiwan	[12]
Cwa	22.40	14.11	Hong Kong	Hong Kong	[5, 8, 11, 27, 28, 70, 71, 77]
Cwa	23.13	113.26	Guangzhou	China	[42]
Cwa	24.15	120.67	Taichung	Taiwan	[12]
Cwa	24.88	102.83	Kunming	China	[42]
Cfa	25.03	121.56	Taipei	Taiwan	[12, 73]
Cfa	−27.67	−48.55	Florianopolis	Brazil	[55]
Cfa	28.23	112.94	Changsha	China	[42]
Cfa	29.76	−95.37	Houston	USA	[67, 68]
Cfa	31.23	121.47	Shanghai	China	[27]
Cfa	33.75	−84.39	Atlanta	USA	[21]
Cfa	35.15	−90.05	Memphis	USA	[67, 68]
Cfa	39.29	−76.61	Baltimore	USA	[67, 68]
Cfa	40.78	−73.88	New York	USA	[58]
Cfa	42.37	−71.02	Boston	USA	[60]
Cfa	44.79	20.45	Belgrade	Serbia	[9, 51]
Cfa	45.07	7.69	Turin	Italy	[10, 14]
Cfa	45.46	9.19	Milan	Italy	[6, 54]
Cfa	45.65	13.75	Trieste	Italy	[46–49]
Cfb	49.28	−123.12	Vancouver	Canada	[67, 68]
Cfb	50.11	8.68	Frankfurt	Germany	[54]
Cfb	51.11	17.04	Wroclaw	Poland	[52]
Cfb	51.51	0.13	London	UK	[44]
Cfb	59.35	18.07	Stockholm	Sweden	[16]
Dwa	37.57	126.98	Seoul	South Korea	[13, 76]
Dwa	45.80	126.54	Harbin	China	[42]
Dfa	40.43	−86.91	West Lafayette	USA	[57]
Dfa	41.88	−87.63	Chicago	USA	[35, 36, 38, 66–68]
Dfa	44.88	−93.21	Minneapolis	USA	[37]
Dfb	44.47	−73.21	Burlington	USA	[67, 68]
Dfb	46.79	−92.10	Duluth	USA	[67, 68]
Dfc	40.25	−105.82	Grand Lake (CO)	USA	[30]
Dsb	64.84	−147.72	Fairbanks	USA	[67, 68]
Dsc	60.72	−135.06	Whitehorse	Canada	[44]
Dsc	61.18	−150.00	Anchorage	USA	[60]

References

1. Aguilar A, Alonso C, Coch H, Serra R (2011) Solar radiation and architectural design in Barcelona: reconciling protection in summer and gain in winter. In: Bodart M, Evrard A (eds) Proceedings of PLEA 2011—the 27th conference on passive and low energy architecture, Louvain-la-Neuve, Belgium. Presses universitaires de Louvain, Louvain-la-Neuve, pp 59–64. Accessed 13–15 Jul 2011
2. Alam MJ, Islam MA (2017) Effect of external shading and window glazing on energy consumption of buildings in Bangladesh. Adv Build Energ Res 11:180–192
3. Ali AAMM (2013) Using simulation for studying the influence of horizontal shading device protrusion on the thermal performance of spaces in residential buildings. Alexandria Eng J 52:787–796
4. Amer M, Wagdy A (2016) Multivariable optimization for zero over-lit shading devices in hot climate. In: Proceedings of the 3rd IBPSA-England conference BSO 2016, Newcastle, UK.

Accessed 12–14 Sep 2016

5. Antoniou A, Yannas S (2017) The residential balcony in the Mediterranean climates. In: Brotas L, Roaf S, Nicol F (eds) Proceedings of the 33rd PLEA international conference "Design to thrive", Edinburgh, Scotland. Network for comfort and energy use in buildings, pp. 981–988. Accessed 2–5 Jul 2017
6. Bellia L, Falco FD, Minichiello F (2013) Effects of solar shading devices on energy requirements of standalone office buildings for Italian climates. Appl Therm Eng 54:190–201
7. Bellia L, Marino C, Minichiello F, Pedace A (2014) An overview of solar shading systems for buildings. Energ Procedia 62:309–317
8. Bojić M (2006) Application of overhangs and side fins to high-rise residential buildings in Hong Kong. Civ Eng Environ Syst 23:271–285
9. Bojić M, Cvetković D, Bojić Lj (2017) Optimization of geometry of horizontal roof overhangs during a summer season. Energ Efficien 10:41–54
10. Cascone Y, Corrado V, Serra V (2011) Development of a software tool for the evaluation of the shading factor under complex boundary conditions. In: Proceedings of building simulation 2011: 12th conference of IBPSA, Sydney, Australia. IBPSA, pp 2269–2276. Accessed 14–16 Nov 2011
11. Chan ALS, Chow TT (2010) Investigation of energy performance and energy payback period of application of balcony for residential apartment in Hong Kong. Energ Build 42:2400–2405
12. Cheng CL, Liao LM, Chou CP (2013) A study of summarized correlation with shading performance for horizontal shading devices in Taiwan. Sol Energy 90:1–16
13. Cho J, Yoo C, Kim Y (2014) Viability of exterior shading devices for high-rise residential buildings: case study for cooling energy saving and economic feasibility analysis. Energ Buildings 82:771–785
14. Corrado V, Serra V, Vosilla A (2004) Performance analysis of external shading devices. In: Proceedings of Plea2004—the 21st conference on passive and low energy architecture, Eindhoven, The Netherlands. Accessed 19–22 Sep 2004
15. David M, Donn M, Garde F, Lenoir A (2011) Assessment of the thermal and visual efficiency of solar shades. Build Environ 46:1489–1496
16. Dubois MC (2001) A simple chart to design shading devices considering the window solar angle dependent properties. In: Furbo S (ed) Proceedings of EUROSUN 2000: the 3rd ISES Europe solar congres, Copenhagen, Denmark. ISES, Freiburg. Accessed 19–22 Jun 2000
17. Ebrahimpour A, Maerefat M (2011) Application of advanced glazing and overhangs in residential buildings. Energ Convers Manage 52:212–219
18. Eskandari H, Saedvandi M, Mahdavinejad M (2018) The impact of Iwan as a traditional shading device on the building energy consumption. Buildings 8:3
19. Esquivias PM, Munoz CM, Acosta I, Moreno D, Navarro J (2016) Climate-based daylight analysis of fixed shading devices in an open-plan office. Lighting Res Technol 48(2):205–220
20. Fadzil SFS, Sia SJ (2003) Recommendations for horizontal shading depths for vertical building facades in the tropic region with particular reference to Penang. Malaysia. Archit Sci Rev 46:375–381
21. Fang Y, Cho SY (2017) Building geometry optimization with integrated daylighting and energy simulation. In: Barnaby CS, Wetter M (eds) Proceedings of building simulation 2017: the 15th international IBPSA conference, San Francisco, USA, pp 2591–2598. Accessed 7–9 Aug 2017
22. Garg SN, Bansal NK (1986) Calculation of appropriate size fixed sunshade overhangs over windows of different orientations. Energ Convers Manage 26:283–288
23. Ghosh A, Neogi S (2018) Effect of fenestration geometrical factors on building energy consumption and performance evaluation of a new external solar shading device in warm and humid climatic condition. Sol Energy 169:94–104
24. Gómez-Muñoz VM, Porta-Gándara MA (2003) Simplified architectural method for the solar control optimization of awnings and external walls in houses in hot and dry climates. Renew Energ 28:111–128
25. Gupta R, Ralegaonkar RV (2006) New static sunshade design for energy-efficient buildings. J Energ Eng 132:27–36

26. Hashemi A, Khatami N (2017) Effects of solar shading on thermal comfort in low-income tropical housing. Energ Procedia 111:235–244

27. He Q, Ng T (2017) Applicability of overhangs for energy saving in existing high-rise housing in different climates. Int J Civ Environ Eng 11:145–151

28. Huang Y, Niu Jl, Chung Tm (2012) Energy and carbon emission payback analysis for energy-efficient retrofitting in buildings—Overhang shading option. Energ Build 44: 94–103

29. Idchabani R, Ganaoui ME, Sick F (2017) Analysis of exterior shading by overhangs and fins in hot climate. Energ Procedia 139:379–384

30. Jones RE Jr (1980) Effects of overhang shading of windows having arbitrary azimuth. Sol Energy 24:305–312

31. Kabre C (1999) Winshade: a computer design tool for solar control. Build Environ 34:263–274

32. Khoroshiltseva M, Slanzi D, Poli I (2016) A Pareto-based multi-objective optimization algorithm to design energy-efficient shading devices. Appl Energ 184:1400–1410

33. Koç SG, Kalfa SM (2021) The effects of shading devices on office building energy performance in Mediterranean climate regions. J Build Eng 44:102653

34. Kottek M, Grieser J, Beck C, Rudolf B, Rubel F (2006) World map of the Köppen-Geiger climate classification updated. Meteorol Z 15(3):259–263. https://www.schweizerbart.de/papers/metz/list/15#issue3. Accessed Mar 2022

35. Krarti M (2021) Evaluation of energy performance of dynamic overhang systems for US residential buildings. Energ Build 234:110699

36. Krarti M (2021) Impact of PV integrated rotating overhangs for US residential buildings. Renew Energ 174:835–849

37. Krarti M (2021) Performance of PV integrated dynamic overhangs applied to US homes. Energy 230:120843

38. Krarti M (2021) Evaluation of PV integrated sliding-rotating overhangs for US apartment buildings. Appl Energ 293:116942

39. Lau AKK, Salleh E, Lim CH, Sulaiman MY (2016) Potential of shading devices and glazing configurations on cooling energy savings for high-rise office buildings in hot-humid climates: the case of Malaysia. Int J Sustain Built Environ 5:387–399

40. Lenoir A, Cory S, Donn M, Garde F (2013) Optimisation methodology for the design of solar shading for thermal and visual comfort in tropical climates. In: Wurtz E (ed) Proceedings of building simulation 2013: the 13th international conference of IBPSA, Chambery, France. IBPSA, Toronto, pp. 3086–3095. Accessed 26–28 Aug 2013

41. León ÁL, Dominguez S, Campano MA, Ramírez-Balas C (2012) Reducing the energy demand of multi-dwelling units in a Mediterranean climate using solar protection elements. Energies 5:3398–3424

42. Li X, Peng JQ, Li NP, Wang M, Wang CL (2017) Study on optimum tilt angles of photovoltaic shading systems in different climatic regions of China. Procedia Eng 205:1157–1164

43. Loonen RCGM, Hensen JLM (2013) Dynamic sensitivity analysis for performance-based building design and operation. In: Wurtz E (ed) Proceedings of building simulation 2013: the 13th international conference of IBPSA, Chambery, France. IBPSA, Toronto, pp 299–305. Accessed 26–28 Aug 2013

44. Maestre IR, Blázquez JLF, Gallero FJG, Cubillas PR (2015) Influence of selected solar positions for shading device calculations in building energy performance simulations. Energ Build 101:144–152

45. Mandalaki M, Zervas K, Tsoutsos T, Vazakas A (2012) Assessment of fixed shading devices with integrated PV for efficient energy use. Sol Energy 86:2561–2575

46. Manzan M (2014) Genetic optimization of external fixed shading devices. Energ Build 72:431–440

47. Manzan M, Clarich A (2017) FAST energy and daylight optimization of an office with fixed and movable shading devices. Build Environ 113:175–184

48. Manzan M, Padovan R (2015) Multi-criteria energy and daylighting optimization for an office with fixed and movable shading devices. Adv Build Energ Res 9:238–252

49. Manzan M, Pinto F (2009) Genetic optimization of external shading devices. In: Proceedings of building simulation 2009: the 11th international IBPSA conference, Glasgow, Scotland. IBPSA, pp 180–187. Accessed 27–30 Jul 2009

50. Nikolaou T, Stavrakakis G, Skias I, Koloktsa D (2007) Contribution of shading in improving the energy performance of buildings. In: Proceedings of the 2nd PALENC conference and the 28th AIVC conference on building low energy cooling and advanced ventilation technologies in the 21st century, Crete, Greece, pp 718–722. Accessed 27–29 Sep 2007

51. Nikolić D, Djordjević S, Skerlić J, Radulović J (2020) Energy analyses of serbian buildings with horizontal overhangs: a case study. Energies 13:4577

52. Nowak H, Nowak L, Sliwińska E (2016) The impact of different solar passive systems on energy saving in public buildings and occupants' thermal and visual comfort. J Build Phys 40:177–197

53. Paydar MA (2020) Optimum design of building integrated PV module as a movable shading device. Sustain Cities Soc 62:102368

54. Panteli C, Kylili A, Stasiuliene L, Seduikyte L, Fokaides PA (2018) A framework for building overhang design using building information modeling and life cycle assessment. J Build Eng 20:248–255

55. Queiroz N, Westphal FS, Pereira FOR (2020) A performance-based design validation study on EnergyPlus for daylighting analysis. Sol Energy 183:107088

56. Raeissi S, Taheri M (1998) Optimum overhang dimensions for energy saving. Build Environ 33:293–302

57. Rao S, Tzempelikos A (2010) The impact of exterior overhangs on the daylighting performance of office spaces. In: Proceedings of the 1st international high performance building conference, West Lafayette, USA. Accessed 12–15 Jul 2010

58. Rocha APdA, Goffart J, Houben L, Mendes N (2016) On the uncertainty assessment of incident direct solar radiation on building facades due to shading devices. Energ Build 133:295–304

59. Sameti M, Jokar MA (2017) Numerical modelling and optimization of the finite-length overhang for passive solar space heating. Intell Build Int 9:204–221

60. Sargent JA, Niemasz J, Reinhart CF (2011) Shaderade: combining Rhinoceros and Energyplus for the design of static exterior shading devices. In: Proceedings of building simulation 2011: the 12th international IBPSA conference, Sydney, Australia, 2011 Nov 14–16. IBPSA, pp 310–317

61. Sghiouri H, Mezrhab A, Karkri M, Naji H (2018) Shading devices optimization to enhance thermal comfort and energy performance of a residential building in Morocco. J Build Eng 18:292–302

62. Shafavi NS, Tahsildoost M, Zomorodian ZS (2020) Investigation of illuminance-based metrics in predicting occupants' visual comfort (case study: architecture design studios). Sol Energy 197:111–125

63. Sherif A, Zafarany AE (2011) Designing the window to fit a shading device: a reversed method for optimizing energy efficient fenestration. In: Leclercq P, Heylighen A, Martin G (eds) Proceedings of CAAD futures 2011: designing together, Liège, Belgium, 2011 Jul 4–8. Les Éditions de l'Université de Liège, Liège, pp 383–399

64. Sherif A, Zafarany AE, Arafa R (2014) Energy efficient shading strategies for windows of hospital ICUs in the desert. Int J Comput System Eng 8:372–375

65. Stevanović S (2016) Search for the optimal shape of fixed exterior shading in a cooling-dominated climate. In: Kitanovski A, Poredoš A (eds) Proceedings of ECOS 2016: the 29th international conference on efficiency, cost, optimization, simulation and environmental impact of energy systems, Portorož, Slovenia, 2016 Jun 19–23. Faculty of Mechanical Engineering, University of Ljubljana, Ljubljana

66. Stevanović S, Naumoska VE, Ralev M (2016) Assessment of optimal shape of exterior shading for offices in hot summer continental climate. In: Tepavčević B, Stojaković V (eds) Proceedings of the 4th international regional eCAADe workshop "Between computational models and performative capacities", Novi Sad, Serbia, 2016 May 19–20. eCAADe and Faculty of Technical Sciences, University of Novi Sad, Novi Sad, pp 10–23

67. Stevanović S, Stevanović D (2018) Optimisation of curvilinear external shading of windows in cellular offices. PLoS ONE 13:e0203575
68. Stevanović S, Stevanović D, Dehmer M (2019) On optimal and near-optimal shapes of external shading in apartment buildings. PLoS ONE 14:e0212710
69. Subhashini S, Thirumaran K (2018) A passive design solution to enhance thermal comfort in an educational building in the warm humid climatic zone of Madurai. J Build Eng 18:395–407
70. Sun LL, Lu L, Yang HX (2012) Optimum design of shading-type building-integrated photovoltaic claddings with different surface azimuth angles. Appl Energ 90:233–240
71. Sun LL, Yang HX (2010) Impacts of the shading-type building-integrated photovoltaic claddings on electricity generation and cooling load component through shaded windows. Energ Build 42:455–460
72. Tahbaz M (2012) Primary stage of solar energy use in architecture–Shadow control. J Cent South Univ 19:755–763
73. Valladares-Rendón LG, Lo SL (2014) Passive shading strategies to reduce outdoor insolation and indoor cooling loads by using overhang devices on a building. Build Simul-China 8:671–681
74. Wong NH, Tong SS, Tan E, Wen JX, Goh A, Lee SF, Li RX (2017) Building façade design for indoor air temperature and cooling load reduction. In: Brotas L, Roaf S, Nicol F (eds) Proceedings of the 33rd PLEA international conference "Design to thrive", Edinburgh, Scotland, 2017 Jul 2–5. Network for comfort and energy use in buildings, pp 891–898
75. Xue P, Li Q, Xie JC, Zhao MJ, Liu JP (2019) Optimization of window-to-wall ratio with sunshades in China low latitude region considering daylighting and energy saving requirements. Appl Energ 233–234:62–70
76. Yoo SH, Lee ET, Lee JK (1998) Building integrated photovoltaics: a Korean case study. Sol Energy 64:151–161
77. Zhang W, Lu L, Peng JQ (2017) Evaluation of potential benefits of solar photovoltaic shadings in Hong Kong. Energy 137:1152–1158